ETHICS, TECHNOLOGY AND MEDICINE

Ethics, Technology and Medicine

Edited by

DAVID BRAINE
Department of Philosophy
University of Aberdeen

and

HARRY LESSER
Department of Philosophy
University of Manchester

Avebury

Aldershot · Brookfield USA · Hong Kong · Singapore · Sydney

Published by

Avebury

Gower Publishing Company Limited
Gower House,
Croft Road
Aldershot
Hants GU11 3HR
England

Gower Publishing Company
Old Post Road
Brookfield
Vermont 05036
USA

British Library Cataloguing in Publication Data
Ethics, technology and medicine.
  1. Medicine. Ethical aspects
  I. Braine, David II. Lesser, Harry, 1943-
  174'.2

LIBRARY OF CONGRESS
Library of Congress Cataloging-in-Publication Data

Ethics, technology, and medicine / edited by David Braine, Harry
  Lesser.
        p.   cm. -- (Avebury series in philosophy)
    ISBN 0-566-05249-0 : £18.50 ($37.00 U.S.)
    1. Ethics. 2. Technology--Moral and ethical aspects. 3. Medical
  ethics.   I. Braine, David. II. Lesser, Harry. III. Series.
  BJ1012.E8955 1988
    174--dc19   88-10519
                                                              CIP

ISBN   0   566 05249   0

Printed and bound in Great Britain by
Athanaeum Press Limited, Newcastle-upon-Tyne

Contents

Contributors

David Braine   is a Lecturer in Philosophy and Gifford Fellow at the University of Aberdeen, and the author of *The reality of time and the existence of God* (OUP, 1988), *Medical ethics and human life* (Palladio Press, Aberdeen, 1983) and numerous articles.

Ruth Chadwick   is the Laura Ashley Fellow in the Philosophy of Health Care at University College, Cardiff, and the editor of *Ethics, reproduction and genetic control* (1987).

Heather Draper   lectures in Philosophy at Crewe and Alsager College of Higher Education and is a tutor for the Open University, and has recently been awarded her Ph.D., in the field of medical ethics, by the University of Manchester.

Simon Glynn   is currently lecturing in Philosophy at the University of Central Michigan, and is editor of and contributor to the collections *European philosophy and the human and social sciences* (1986) and *Sartre: an investigation of some major themes* (1987).

John Hostler   is Staff Tutor in Philosophy in the Extra−mural Department of the University of Manchester, and the author of *Leibniz's moral philosophy* and of several articles in the field of ethics.

David Lamb   is a Senior Lecturer in Philosophy at the University of Manchester, author of *Death, braindeath and ethics*, co−author of *Multiple discovery: the pattern of scientific development*, and editor of the collection *Hegel today* and the journals *Explorations in Knowledge* and *Explorations in Medicine*.

Harry Lesser is a Lecturer in Philosophy at the University of Manchester, co—author of *Political Philosophy and Social Welfare* (RKP, 1980), and author of several articles in the *Journal of Medical Ethics* and elsewhere.

David Linton was, until his untimely death, a post—graduate student in philosophy at Balliol College, Oxford.

David Smail is District Psychologist for the Nottingham Health Authority, and the author of *Illusion and reality: the meaning of anxiety* (Dent, 1984) and *Taking Care: an alternative to therapy* (Dent, 1987).

Acknowledgements

The editors would like to express their deepest gratitude to all the contributors, and to all those who have made helpful comments and criticisms on early versions of the various papers, in particular those who attended the conferences of *Ethical Issues in Caring* at Manchester in September, 1984 and 1985, when the papers by Heather Draper and Harry Lesser were read; and of the *Northern Association for Philosophy* at Manchester in January, 1982, when the papers by Ruth Chadwick, John Hostler and David Lamb were read.

They would also like to thank the editors of *Explorations in Medicine* (April, 1987) for permission to reprint the paper by David Lamb; and they are particularly grateful to the editors of the *Journal of Value Inquiry* for the opportunity to reprint what is, sadly, the only fruit of the great academic promise of David Linton, first published posthumously in the *Journal of Value Inquiry*, vol.XIII, no.1, Spring, 1979.

Finally, they would like to say how very grateful they are to Liz Brock for her patient typing of a difficult manuscript.

1 Introduction

The papers in this volume constitute an attempt to define and to respond to a group of ethical problems raised by the growth of technology, especially in its application to medicine. We see these problems as interconnected, and the basic ones as stemming not so much from the particular advances in technology as from the growth of the technological attitude. Technology itself has obviously brought great benefits; but it has also brought its dangers. We do not in any way belittle the benefits: but the concern of the authors of this book is with the dangers and limitations of the approach to human beings inspired by this technological attitude. Some of the papers deal with this approach in general, and some with specific issues; but these are connected, inasmuch as the views and attitudes being discussed are not held as pure theories but as leading to practical consequences; indeed, much of the time the authors are dealing with attitudes that are, sometimes unconsciously, presupposed by what people actually do, rather than views that they hold explicitly.

The hallmark of the technological attitude or approach lies, it may be said, in seeing the human being as essentially a problem—solving animal. This involves a view of people, of education, of suffering, of the environment and of values, all interconnected. People are important, on this view, because of their superior intelligence to animals, not because of their feelings, emotional responses or physical activities and enjoyments, except where these are instruments of intelligence. Education, if successful, is essentially the learning of various skills which harness this intelligence in ways which improve the conditions of human life, chiefly the material conditions. Suffering, and difficulties and problems in general, are to be dealt with by diagnosing the specific cause of the problem and intervening to try to alter and improve it. ("Diagnosis" is being used here in a general sense.) The natural environment, including the non—human animals, is to be manipulated to human advantage, or supposed advantage. Finally, values arise simply from human choices, whether made by individuals or agreed by groups; and human skills, intellectual and physical, are to be used to bring about the ends set by these choices, and to remove obstacles in the way of this. Probably no one in practice holds these views to this degree all the time and in every area of life: indeed, it may well be psychologically impossible to do so. The attitude, moreover, is, when held appropriately, i.e. when dealing with problems that are essentially technical, extremely useful and beneficial. But at the present time it is often applied inappropriately and too widely; and this is when it does harm rather than good.

First of all, even as a problem—solving attitude, it is appropriate only to those problems which are relatively isolatable, and do not require a "holistic" approach, i.e. an approach that considers the whole person, at least, and often the person in society and in the natural environment as well — one may contrast, in medicine, the problem of a broken limb and the problem of a reactive depression. More fundamentally, it makes both too much and too little of human intelligence. Too much, because it ignores the limits set on what is possible by our biological nature, and by the fact that we are social animals and both the family and society are essential parts of our humanness. Too little, because human mental powers are expressed not only in technology but also in, for example, art, religion, play, creative work and friendship, none of which can be reduced to problem—solving. More fundamentally still, it is an attitude based on manipulation, and lacking in respect. It fails to respect the environment, and biological life as such, fairly obviously; but ultimately, by failing to see human beings as persons in society, it fails to respect either the human community or the human personality.

It is not only that the technological attitude needs to be supplemented by a more humane and holistic one. The attitude of respect for humanity needs to be prior; the attitude of diagnosing, intervening and manipulating needs to be subordinated to it. Such is the general presumption of this collection, argued for somewhat differently in different articles; but generally agreed. Acceptance of it is independent of any religious view, unless respect for the world around one is called "religious"; indeed the authors include both theists and atheists, and see the arguments here as independent of theology — one might try to argue from what is said here to a theological position, but not the other way round.

The technological attitude is often associated with the view of mind and body called "physicalist", which not only identifies mind and brain, but holds that, at least in principle, human beings can be understood in purely physical terms, without reference to thoughts or feelings as experienced, so that all human problems which are solvable can be solved by the appropriate physical action. It is part of the "humanist" attitude for which we argue that it maintains that there is no way of doing this, and that no proper understanding of humanity is possible without taking the mental into account. But it also opposes another version of this approach, which is based on the contrary, "dualist", assumption that mind and body, though they interact, are independent of each other, and that the body is the mind's instrument. Both these views, it is contended, need to be rejected, if we are to do justice to the facts of our experience; and we must instead accept that a human being is, to use the jargon, a "psychophysical unity", necessarily part both of the natural world and the human social world, who cannot be removed from either.

But, as a final preliminary point, it should be noted that not all forms of technology engender the attitude described, although it may be the case that much present—day technology tends to do so. This means that there are three problems. The first is to describe and justify a response to human nature which does justice to its various aspects as a totality and to the obligation to respect it. The second is to determine, within this response, when and how the technological approach is useful and valuable. The third is to determine what forms and uses of technology, apart from their efficiency for specific purposes, which is a technical and not a philosophical question, will encourage rather than work against this respect for human nature.

David Braine's *Human animality: its relevance to the shape of ethics* addresses the first and third of these questions. It argues, on the one hand, that the good is primarily the object of appropriate rounded enjoyment, and only secondarily, not primarily, as it is often taken to be in contemporary ethical theory, the object of choice or wanting; and on the other hand, that because neither dualism nor physicalism is defensible as an account of human nature, and because we must therefore see human beings as intrinsically psychophysical, ethics must therefore take our physical being seriously, and deal with our whole nature, including what we share with other animals, and not merely with our "personality", narrowly defined as our capacity for conscious choice. A consideration of the question of what objects of enjoyment are appropriate to such a being, it is argued, both shows the error of taking a purely technical approach and provides a standard for judging the rights and wrongs of particular uses of technology, various examples of this being given in the final section.

Simon Glynn's *Objectivity and alienation: towards a hermeneutic of science and technology* makes use of the work of philosophers in the phenomenological and existential tradition of twentieth—century Germany and France, notably Husserl, Heidegger and Sartre, to offer an account of how and why the particular way in which technology has developed in the West in recent centuries has led to the widespread adoption of an alienated and alienating technological attitude, distancing people from their environment and from each other. It goes on to argue that this constitutes not a mere theoretical mistake but a serious and imminent practical danger; and to consider how a change in attitude, together, and interacting with, a change in the kinds of technology we use (not an abandonment of technology as such) could do something to minimize or remove the danger.

Harry Lesser's *Technology and medicine: means and ends* tries to apply similar considerations to medicine in particular. It argues that the great advances in medical technology have altered the attitudes of both doctors and patients, and in particular have led to a great expansion of the areas in which the "medical model" of diagnosis and intervention is both thought appropriate and used in practice. It goes on to argue, with particular reference to psychiatry, that this has brought both benefits and dangers; and that the dangers can best be avoided by seeing medical technology in a wider human context.

David Braine's second paper, *Human life: its secular sacrosanctness*, moves the argument to a more specific level. It argues that an examination of the necessary conditions of human existence, notably its social nature, its being necessarily physical and the unity between the various things that make up a good or worthwhile life, shows both the errors of individualist and utilitarian approaches to ethics, the intrinsic value of mere biological human life, and the necessity of the virtues, such as courage and humility, to a good human life. Hence, it is argued, the taking of human life is not wrong because it tends to do more harm than good, but because it cannot be done without removing an intrinsic good, and violating, for instance, the virtues of fortitude, humility and respect. There is thus an absolute bar on taking human life, including one's own, perhaps removable by a person's actions (such as threatening another's life) but not by anything in their nature or a condition typical of their nature, such as handicap, weakness or age.

John Hostler's *The sanctity of life and the sanctity of death* argues rather similarly for the sanctity of human biological life. It also criticises attempts to argue for this in the first instance from theological, hedonistic or personalist assumptions; and then argues that biological life is valuable as the precondition of everything else valuable. Its value is enhanced, the argument goes on, by the consciousness of death, and the awareness of the need to use the time available and not to waste it, including the time available in the final stages of life. This imposes obligations on medical staff and relatives — to be honest with the dying, to help them understand their situation, and to help them die with dignity. These obligations complicate medical ethics; but this is inevitable once medicine deals, as it should, with people rather than with their bodies alone.

David Lamb's *Down the slippery slope* turns our attention to possible exceptions to the principle of the sanctity of human life, notably abortion and euthanasia. It investigates one particular argument against these practices, namely that to allow them under any circumstances, for however humane reasons, opens the door to a total disregard for human life, since each case resembles the next so closely that no logical stopping—point can be found. The argument is closely examined; and it is contended that it fails as an argument against abortion, because one *could*, on Lamb's view, correctly or incorrectly, maintain that there is a fundamental difference between a fetus and a born human being, but succeeds as a valid argument against the legalisation of euthanasia.

The first three articles thus deal with respect for humanity as a psychophysical unity; the next three with respect for human biological life. All six are concerned to challenge the technological approach when it is at variance with these principles. The next three papers address rather more specific issues involving technology. The first, Ruth Chadwick's *Genetic Improvement*, considers the eugenic use of modern reproductive technology. It considers the arguments against eugenics; and argues that there is no ground for maintaining that eugenics is morally wrong, or for accepting negative and rejecting positive eugenics, except when the method used is morally objectionable on other grounds. The paper does not attempt any full discussion of this issue, but does argue that most current methods, including all the "old", traditional ones, are open to objection. As regards the new methods of modern reproductive technology, she tentatively concludes that *in vitro* fertilization is the one least open to objection, and also that assessment of any method should not be purely individualistic, but must especially consider the effect on society as a whole.

Heather Draper's *Transexuals and Werewolves: the Ethical Acceptability of the Sex—Change Operation* discusses the use of the sex change operation as the primary method of tackling the problems of the transexual patient. It is argued that this is a misuse of medical technology, and the appropriate discussion of the problems of these patients belongs to psychiatry and not to surgery.

David Linton's *Why is pornography offensive?* breaks the run of papers on medical issues. But it is connected with the general theme of the book, because the "dehumanizing" quality of pornography seems bound up with its replacement of a personal attitude to sex by a technological one. The paper investigates the source of the feeling that pornography, "the treatment of sexual practices divorced from any tender consideration" offends and degrades, and argues that it is because it is perceived as a symbolic threat to society's integrity. A further conclusion is that it is the business of the law not to try to eliminate pornography as such, but to try to protect people from having obscene material thrust on them.

Finally, David Smail's *Technology and Psychotherapy* returns to the general issue of the technological attitude, as applied in the practice of psychotherapy. The paper discusses the reasons for the attitude's being so widespread among psychotherapists and often the public at large; and the reasons why it is doomed to failure, since people are not machines that can be repaired or adjusted when they go wrong. Dr Smail's conclusion, which may serve as a conclusion to the whole book, is that "(a)voidance of .... psychological damage will not be achieved by technological means, but only through our developing a society in which we treat each other with greater care and kindness" — a care and kindness which, as is shown in the arguments of the papers in this book, has to be informed by a consciousness of man as a psychosomatic unity and an animal set within an environment, and by respect for what is in this way given by nature.

2 Human animality: its relevance to the shape of ethics

DAVID BRAINE

My thesis in this paper is that the only way of obtaining a right approach to the ethical problems in regard to technology and its right use is through grasping how integral man's animalhood is to his humanity and how his animality shapes "the good for man".

I shall argue that the good is primarily the object of appropriate enjoyment, and only derivatively from this the object of wanting and seeking. Correspondingly, what is wrong is that which is systematically or structurally destructive of a rounded enjoyment: a rounded enjoyment or happiness is not a sum of bits and pieces but a structured integrated whole [1], and, it can be argued, a whole that embraces the entire human community, proper and fulfilling relationships and attitudes within which are integral to every individual's happiness. In considering the good for man, the proper starting point is to identify the objects of enjoyment appropriate to man and the structured relations between the different aspects of life involved in full human living. It will become clear that human enjoyment, human happiness and integrated human life can have a coherent account given of them only if the intrinsic value of the psycho−physical is admitted, i.e. if man is throughout treated as a bodily being who in the same acts fulfils and in the same acts violates at one and the same time his personhood, his animalhood and his humanity. Indeed, the intimacy and the extent of the given or non−voluntary character of the communal or person−to−person values and obligations between human beings is, it can be argued, a result of the way in which communality is built into human nature in a many−levelled and all−embracing way precisely and only by his character as an animal, that is as a being rooted and set within the system and community of family, mankind and the whole environment of living beings and nature.

I shall begin by sketching that view, diametrically opposed to this, which is currently culturally dominant, the view most conveniently labelled 'personalistic', typical alike of utilitarians, of followers of Kant and Sartrean existentialists, of dualists and mind−brain identity materialists, of Calvinist

Protestants and modernist Roman Catholics. And I shall indicate why the two main roots outside ethics of this standpoint have no claim to be considered determinative of method in ethics. Rather, unless and until some demonstration has been produced of these alien suppositions, ethical inquiry should proceed in a way appropriate to its subject—matter, exploring what belongs to the good for man and what is contrary to it, in a way compatible with the general lineaments of the concept of good.

These preliminaries will leave us free to pursue our positive inquiry into human good, thereby to reveal how intimately the personal is interwoven with the biological and bodily, so that it is open to us to violate our personhood and our humanity most deeply in treating our bodies or our environment merely technically. In this way, we may recover a sense of the ethical point (as distinct from the vital, but to most educated western men and women more obscure, metaphysical point) of the Jewish and Christian insistence on the unity of man's body and soul and the need for a resurrection of the body. This latter insistence dovetails rather with the naturalistic character of most primitive religions and ancient mythopoeic thought (which always tended to investigate man and the community of mankind in the context of the community of nature) than with the conception of mind, and the human mind in particular, as transcendent, independent and dominative.

Section 1 : Personalistic Ethics

In common presentations of various viewpoints in modern philosophical ethics, viewpoints in many other respects mutually opposed, it is taken as a sacred datum that ethics must be 'personalistic'. It is not of course a datum for all philosophers, since it is not a datum for the followers of Aristotle, many idealists, those influenced by Wittgenstein or Heidegger, or the many inspired by anti—dualistic or environment—orientated religions or cultural perspectives. But none of these set the dominant tone of contemporary philosophical ethics.

Let us sketch this 'personalistic' standpoint in what might seem its intellectually and humanly most attractive form. It is not possible to bring all the surprisingly kindred views indicated earlier within one undisputedly common description. After all, what utilitarian regards all persons as having the character of 'ends in themselves'? Yet even the utilitarian counts persons as if in some way to make it persons that matter for ethics. Rather our purpose is to portray not a straw man but a recognisable strangely prevalent, indeed dominant, tendency in thought, in order against this backcloth to set up by contrast our own positive exposition.

For the personalist, any appeal to any biological fact about human nature is ruled out of court from the start on grounds that it would involve an invalid inference from fact to value and a failure to recognise the full ambit of the human capacity for critical reflection on any supposed data, even data allegedly integral to human nature as such. We are to treat all persons, and not only ourselves, as having the character of 'ends in themselves', always to be respected as ends and never treated merely as means. And we are to act always as persons, not sheltering behind any role supposedly settled upon us by society, nature, or by our own previous choice or choices, but always experiencing complete responsibility for our

choice and for the amount of self−reflection (holding open to question all alleged roles or other supposed data of deliberation) it issues from. As persons it is imagined that we may take to the limit some road of self−sufficient self−realisation (cf. Nietzsche, or alternatively some road of love, however understood. Which road we may follow is an expression or realisation of our choice, our self−will. Set thus as individuals without law from outside ourselves, it follows that, if any morality whatsoever is to govern us, it will have to be a morality expressive of what we are *qua* persons and *qua* nothing else: it must be a morality which, one and the same, would govern in the same way all rational, intellectual or choosing beings as such (whether bodily beings such
as men, or pure intelligences such as the angels have been supposed to be).

Now indeed, at least *prima facie*, this is a very odd view to take of human morality. The most obvious examples of moral principles have to do with killing or being responsible for the biological death, mutilation or deliberate harming or damaging of another, the infliction of pain or the avoidance of this, and duties associated with the family, to parents, children, siblings or those of the same stock, i.e. other human beings. Further principles seem to relate to the sharing of work, of the means to work and of the products of work, and to the sufficiently reliable possession of these, and of clothes, hearth and shelter, for a measure of well−being in personal and family life to be practicable. And, if custom establishes some limited rights in respect of the giving during life or disposition at death of these things, then there will also be principles marking the respect due to such customs within the limits set by the underlying roles of these things in human life. The care of the dependent, whether the immature, the diseased, the disabled or injured, or the old, is likewise a matter scarcely pertinent to the morality of a community of angels.

If we take some apparently 'personalistic' approach, then we seem trapped into holding that nothing psycho−physical has intrinsic value as such in itself, and into proceeding as if we were superior or transcendent entities or subjects (called "selves" or "persons") who have the duty of making a proper use of human bodies and their parts, as part of a more general duty of administering, so far as it lies within human power, or the human sphere of influence, the physical world as a whole, more particularly the environment close to man.

It now takes a bold man in the circles of liberal or 'western−culture' thinking to attribute any intrinsic value to the physical as such. The notion, or the realisation, that there is (or might be) intrinsic value in physical systems as such in their order and apparent wonder, so that there would be good in their existence, even if man had never existed to behold them, is regarded as discredited in any sober movements of reflection, and the received view − not in fact the view of Keats − is that beauty lies only in the eye of the beholder.

Against this background, the province of value contracts. The things which, if anything, have intrinsic value are various supposedly mental phenomena, whether these include thoughts and structures of thought and will, or whether they be conceived of in terms of pleasure, pleasures being commonly envisaged as if they were a species of feeling or sensation. Alternatively, if the idea of intrinsic value is rejected as involving an untenable objectivism, then the attitudes whereby value is attributed to things are regarded as rooted (in origin, not justification) in evolution or in the sociological development of conventions or of interests, so that ethical judgements are grounded in ultimately arbitrary (though not without cause) decisions of the will. (Of these decisions of the will, "interests" and the "balance of interests", however assessed, are an expression).

Thus, once we have repudiated any intrinsic value in anything psycho−physical, and discarded the intrinsic value of the physical, the sphere of value has contracted to that of the mental only. The disciple of Plato began by being enraptured by the wonders of the objects of theoretical contemplation. His descendant now administers the universe, or what little of it he can control, with a view to the attainment of the ends on which his arbitrary will has settled, or the thrills or sensations which might fall his way, by contrivance or good fortune. It is now a common opinion, associated often with the mind−brain identity theory, that it does not matter whether these 'happy' sensations spring from experience of the outer world (mistakenly supposed to have some intrinsic value) or whether they have been induced by some drug or brain surgeon.

Section 2 : The Roots of the "Personalistic" Misconception

The roots of this very odd ethical standpoint, which has thus become so predominant in the westernised world, seem to be two: a naive conception of the dilemmas of metaphysics and a naive theory of the conditions of rational judgement.

As to the first, it is commonly assumed there is no other alternative than either a physicalistic materialism or a traditional dualism. That is, it is presumed that there is no choice except between, on the one hand, an ultimately materialistic view of man within which, although the mental may be brought in as providing a simplifying system of causal explanation, the fact remains that the only real causation is physical, and, on the other hand, a dualism according to which man includes a non−material part, called a mind or a soul, in addition to his various material parts, interacting with the material parts or having states correlated with theirs. In the physicalistic view, although man may have aspects more immediately known or readily described in mentalistic terms and although these terms may serve to identify some overall structural features in human mental life and behaviour and so in a certain sense have an explanatory role, the ultimately determinative explanation of human behaviour, to the extent that any determinative explanation exists, is exclusively in terms of phyusical events, and physical relations and laws.

On both these views, it is assumed that, in any account of man, reference to the human being as such (one might say 'as a whole') can be eliminated in favour of reference to his parts and their inter−relations. Typically, at the first stage of analysis they distinguish an inner part, 'the inner man' − whether this be conceived of as a non−material entity (a mind or soul) or whether it be identified with the brain − and an outer part, the aggregate of man's bodily parts or of the rest of these parts apart from the brain, 'the outer man' whose behaviour we observe in speech and action. The inner man or mind (whether or not identified with the brain) is then being conceived of as governing and using the outer man, as a mind governing, using or administering a machine (and through it the environment of the machine).

In this way dualists and mind/brain identity theorists are forced by the isomorphism of their metaphysical views into a sameness of ethical standpoint. The possibility that both these types of metaphysical view may involve precisely the same reductionist mistake, of regarding wholes as having their nature and behaviour explicable in terms of the nature, behaviour and inter—relations of their parts, and in this way reducible to their parts, is left out of consideration. Yet this neglected holistic view is not only the most immediately natural pre—critical view, but also represents the perspective forced upon one by the argument of such philosophers as Ryle, Wittgenstein and Strawson, together with a massive company of phenomenologists and, before them, in much earlier times, Aristotle and Aquinas.[2]

More precisely stated, the holistic view to which I refer is the following. Human beings and higher animals see, desire, imagine, remember and act by desire, in such wise that reference to them as logically unitary subjects of these activities, i.e. reference to them as wholes rather than as simply heaps or physically inter—relating aggregates of parts, is not to be eliminated from the description and explanation of their behaviour, and indeed not to be eliminated even from the description and explanation of the behaviour of their bodily parts. The nature of these logical subjects is indeed such as to involve their having bodily properties and therefore such as to involve their having bodily parts: they are indeed wholes constituted with parts. And part of the meaning of the description of these parts as bodily is that much of their character and behaviour can be considered in abstraction from the wholes they make up: e.g., animals have physical parts, some aspects of whose behaviour are considered in physics, chemistry, etc.; and such consideration is often of key importance in the causal explanation of malfunction in the larger wholes they enter into. Yet, these parts remain such as to have their overall nature — i.e. the principles of their behaviour when this is considered overall — intelligible globally, as distinct from merely in certain limited aspects, only in functional relation to the wholes they make up. In still describing the finger cut off in battle as a 'finger' we attend only to its historical origin, and abstract entirely from the functional relation, of being a functioning living part of a living human being, which entered into the explanation of its nature and behaviour before being cut off, and gave point to describing it as a 'finger' at all.

To establish any such holistic perspective in metaphysics is not the task of this paper. But since the rival perspectives, respectively dualistic and materialistic, have no established position, but rather, if anything, call for outright rejection on overpowering metaphysical and epistemological grounds, it is not incumbent on the moral philosopher to take these strange rival, dualistic or materialistic, views as data. Rather the moral philosopher is free to seek out whatever approach seems to accord with what his subject—matter, viz. human life and value, invites.

As to the second root of the 'personalistic' ethics which I described, this lies in a certain naive theory of the conditions of rational or critical judgement; thanks above all to Descartes and Kant, it has come to be supposed that the nature of human freedom, just as much as the supposed nature of human thought and reflection, requires an ego which is in no way internally limited by its nature or by anything empirical, so that no other conditions are set upon human deliberation except ones which are purely formal. Descartes raises the epistemological question as to the conditions of knowledge and deliberation as to what is true, while Kant implicitly broadened the inquiry from mere epistemology to the problematic of judgement as such, including practical judgement. In this way, the transcendental ego of Kant can come to be envisaged as involved in any exercise of freedom as well as in a theoretical inquiry and self—reflection. In what Sartre draws from his Cartesianism, the conception of an existential freedom is the most prominent immediate consequence or mark of self—consciousness, and in this way of thinking the conditions of authenticity are of more interest than the conditions of theoretical knowledge. Indeed the picture we have just drawn with a few bold strokes of the brush raises the question as to whether the supposed difficulty in ever deriving an "ought" from an "is" may not stem from this same all dominating perspective wherein one is trapped into viewing an untrammelled Subject as set over against a World within which somehow this Subject has yet to come to know and to operate.

However, there is no good reason for ethics to proceed as if this crude, Subject—Object, problematic of judgement had to stand as a datum, or even as a reasonable presumption.

We are ourselves persons, or, if the word must be used, "Egos", and we meet with, and have some knowledge of, other persons or Egos, bodily beings like ourselves, together with us in the same world. *Prima facie*, these persons, or to be more precise these human beings, have a nature in some sense — for instance, it is an aspect of their nature to have the capacity or capacities required for the competent use of language —
and, *prima facie*, they do exercise free choice and some capacity to reflect on themselves and their own procedure of thought. It is only in the exercise of human freedom and critical self—reflection that one can in any disciplined, rigorous or analytical way discern in what ways the nature of human beings might function or misfunction in such a way as to limit this freedom in choice and this underlying freedom in self—reflection.

However the nature of human beings is such as to allow, through the use of language, the possibility of the posing of alternatives and of deliberation. And it is also such, through growth within the context of interrelationships within a human community, as to make possible the apprehension of the intrinsic value of a variety of goals, and the variety of ways in which these may be structured together in a happy or fortunate life, and thereby allow the individual to mature so as to become responsible for his own judgements and choices. The structures of human language and human community, in large part empirically knowable, will therefore also be preconditions of a mature human being's capacity for freedom and self—reflection.

Accordingly the view which, if any, should be presumed methodologically valid, until proven to be otherwise, is the view that various empirically accessible aspects of human nature are rather preconditions of, than obstacles to, maturity and authenticity in the exercise of freedom and the establishment of knowledge. In this way, consideration of the problematic of judgement reinforces a certain holism, and does nothing to recommend either dualism in metaphysics or the much vaunted "personalism" in ethics which I am in this paper occupied in attacking.

Section 3 : The Notion of "Good"

In considering what is good for man rather than evil, right rather than wrong, modern moral philosophers have tended to give primacy to a consideration of how one is to decide about the rightness and wrongness of actions. Next, considering actions as issuing from desires, they have attended to the notions of desire and interest, and considered 'good', or 'the good', as primarily the object of desire or wanting. As against this I argue that good is not only the object of desire (the desire is for that which we do not have) but also the object of enjoyment or pleasure (enjoyment is *of*, and pleasure is *in*, that which we have) and that it is this relation to enjoyment which is primary for the understanding of the notion of good. This will lead us in Section 4 to seek out the characteristic objects of man's enjoyment, i.e. to identify what it is appropriate for him to find happiness in appreciating, contemplating or enjoying.

Accordingly, I need to begin by making certain logical observations about the priority of the relation of good to enjoyment, and the distortions which result from ignoring this: then some remarks on the ambiguities in the notion of "having" when we speak of enjoying the goods we 'have', and the relation of these distinctions to the traditional distinction between the intrinsically good, the advantageous or useful, and the pleasant; and some remarks on how the very notion of enjoyment opens the way to considering human happiness as a structured whole in which different types of enjoyment, and enjoyment of the same thing or activities at the same time in different respects, are integrated together.

Good as the object of value [3] is not only the object of wanting and seeking, as when we try to get things which we do not have, but also the object of the satisfaction, enjoyment and pleasure which we obtain in that which we have already, whether we get it by our own action or whether it exists independently of our action.

Logical confusion has entered deep into modern moral philosophy because in it good is considered primarily as the object of desire or wanting. Rather, logically, good has first to be the object of satisfaction, enjoyment or pleasure, before it can be an object of wanting, seeking or trying to get. Wanting, seeking and acting are directed towards some object in which satisfaction, enjoyment or pleasure will be obtained or in which it is anticipated these will be obtained.

Yet in modern moral philosophy preoccupation has been almost entirely with wanting and with action, wanting being the spring of action, and the action being conceived of as getting or achieving some goal external to the action itself, i.e. some goal considered as a consequence of the action (often pleasure is ranked merely as an effect or consequence). Hence, for instance, the question of the objectivity of values is commonly considered in terms of the objective goodness or badness, worth or lack of worth, of proposed goals in which enjoyment or pleasure might be anticipated.[4]

This preoccupation with action shows itself also in a distortion of the consideration of the dimensions of human freedom, virtue and vice.

It is plain that there is some moral value, good or bad, involving freedom, associated with the distinctions between wisdom and folly, hope and despair, trust and mistrust, and with *accidie* (mistranslated "sloth": it is in ancient traditions a vice whose symptoms are as much restlessness as laziness: "boredom" would be nearer the mark as a translation). In all these things freedom is in some way expressed but choice is only indirectly involved. Therefore the notion of freedom of the will in man is much wider and richer than the notion of free choice.

However, if we avoid this preoccupation with action, and avoid this tendency to make the consideration of desire prior to the consideration of pleasure or enjoyment, we should note the variety of things that we do take pleasure in or enjoy. Many of them are not products of our will at all; the starry heavens, the sea, landscapes and so forth, and most certainly God. This will be a key point in our later argument.

It is a mistake, in considering the distinction between, on the one hand, wanting things that we do not have, and, on the other, enjoying, being pleased at, or rejoicing in, the things we do have, to be guided by the irrelevantly wide use of the word "have". For some things are said to be 'had' only in the logical sense that they are attributes of things, i.e. characteristics it is good to exemplify, things perhaps it is good to be; it is good to be faithful and caring and courageous and hopeful, serene, lively, strong, agile, vigorous and perceptive. Others are 'had' as tools and means; refrigerators, cars, bicycles are in our possession in some customary or quasi—legal sense.

In all thinking about the word "good" and its variety of use, it is important to keep clear the traditional distinction between the intrinsically good, the useful or advantageous, and the pleasant (i.e. that which we enjoy or are pleased at or in). "Pleasant" in this sense covers the whole range of things which are good under the aspect of the enjoyability of knowing them appreciatively, whether the knowledge be more intellectual or less so, whether it be through the elegance of proofs, or through the elegance revealed in paintings.

So much for general initial remarks about the word "good". If we consider human good, and what is involved in being a fortunate person and in that sense a 'happy' man or woman, it is clear that it is not just contentment but rather consists in a certain structure. It appears that any individual will be unfortunate, and in that sense unhappy, if he misses out on any of four central areas of human fulfilment. First, there is appreciative knowledge, as mentioned above. Second, there is creativity and production, whether in art, craft or nurture of family or society. Then, thirdly, there are personal relationships, whether in family, with friends or otherwise. And I am inclined to add a fourth, namely play. However, this last turns out, as is exhibited in Hugo Rahner's book *Man at Play* [5], to be exceedingly intimately tied up with the capacity of contemplation and wonder as well as with joy in adventurous creativity, and that is the reason for my uncertainty as to whether it would be listed separately. The properly structured human life — by this I mean the human life structured in a way which is conducive to or constitutes a happy life, a fortunate life — involves all these four in some kind of integration.

Creativity, whether instanced in building and carpentry, skilled handicraft or works of art, in the procreation, nurture and education of children, in the nurture of society by those who, if rightly motivated, have care for its good and health, as well as the power and responsibility for this (viz. governments), or in the creativity involved in any episode of intrinsically good activity as such, represents the second of the four aspects interwoven within human happiness which I enumerated earlier. There is not space in this paper to develop an extensive discussion of this aspect of human nature. In this paper I must limit myself to observing the inseparability of creativity, i.e. productivity considered as something appropriate to take pleasure in and not merely as a means (let alone merely a means to the production of external goods with only the status of means), from the others. It is inseparable from play, from appreciative knowledge sometimes expressed in play, and from personal relationships where, associated with appreciative knowledge, exercises of creativity represent the chief thing in which friends and lovers share.

Section 4 : The Objects of Human Enjoyment

I have made these remarks about the analysis of the concept of "good" and the primacy of enjoyment and contemplation in relation to this notion, because, when it comes to identifying the basic goods of man, if one uses the consideration of what man finds pleasure or enjoyment in, an objection may arise of the following kind. Somebody might say "So what! That is how man evolved, so perhaps it was useful in some way for him to have contemplative tendencies and to find pleasure or enjoyment in these rather than other things; but how does this do anything at all to indicate the intrinsic value either in the pleasurable contemplation or enjoyment of these objects and activities, or in these objects or activities themselves?". And this might represent a valid objection to some of the arguments of Mary Midgley in her book *Beast and Man* [6]. However I take it as a datum that consideration of the good *qua* useful is posterior to the consideration of the goal which it is desirable to attain by means of the useful, i.e. the goals in which pleasure, satisfaction, enjoyment are to be had for their own sake; and it is in this latter place that appreciative knowledge and contemplation have their location. And I take it that appreciative knowledge and its objects, when generically thereby appreciated as pleasing, i.e. as beautiful, are good. One use of the word "beautiful" is precisely to signify the good *qua* pleasing.

Accordingly, my point is not that man is a contemplative animal, but that, granted that contemplative pleasure is going to be integral to the good of any kind of being with experience or knowledge, it is to be noticed that man's contemplative activity is in typical and key cases intrinsically animal in its manner and objects. It is a mistake to justify regarding contemplation or its objects as valuable to man by arguing that contemplation is an element in our kind of animality (some things, but not this, may be justifiable in this way). The value of contemplation and its objects is a presupposition of anything's being worth seeking at all because it is a presupposition of anything's being at all satisfying. Rather the question is of the kinds of contemplation, the kinds of appreciation, the kinds of pleasure appropriate to man.

It is entirely topsy—turvy to develop one's moral philosophy as if the primary point of contemplation of human life was as a means of recreation necessary to health, the health being required for industry in the making of things useful to man. Contemplation may have, and indeed it does have, this role, but logically it is a secondary role: its primary place is in consideration of the ends to which it is desirable to obtain for their own sake.

Amongst the things good to man to contemplate or appreciatively know, chief are various things not constructed by the contemplator but given by nature: the stormy heavens on a cloudless night; skyscapes of such rich cloud—colour configuration; the sea in its strength and power, whether in wild or in calm landscapes; the configuration of objects, still and unmoving but yet some living; the movement of plants in the fierce wind or in their light grace, their fragility, their strength of stance or their solidity and age; grace, skill, agility, firm strength in the stance or movement of animals and human beings, as in dance and ballet. Add to these music, the sonority of certain poetry when read aloud, and song.

All these things have a certain order, discovered by the astronomer, uncovered by the anatomist, exhibited in musical scores and commentaries. But the appreciative wonder at these things does not wait upon the intellectual analysis or description of such order. We wonder at the heavens as an abyss of mystery even while the arrangement of stars still might appear to us rather higgledy—piggledy. The awe of the sea depends little on the science of wave motion. Knowledge of musical science may train us in listening, but distracts if verbalised while listening.

Thus many of man's noblest pleasures are not purely intellectual (the more purely intellectual seem located in mathematical elegance of proof, the niceness of chess gambits, and other such rarefied areas). Now, if man views himself as a mind in a machine, but rejects Platonic and idealist views which see value in the intellectual as such, then *he must represent* these pleasures as valuable only as offering relaxation, i.e. as a *means* to more efficient production at other times, or as valuable only *qua simple pleasures*, offering sensations and thrills which human beings happen to want, but with no value deriving from their having wonder or beauty in themselves. The beauty and wonder of these things is, in this conception, only in the eye or ear of the beholder, and this value of theirs in the eye of the beholder has no more status, in reason, than the taste of a strawberry to the unsophisticated. The sensible pleasure has been isolated as an object, treated as merely physical, with an only accidental or empirically noticed relation to the discrimination or judgment evaluating it. Thus the account given, instead of explaining how these pleasures are appropriate and ennobling to man as a unity, as an animal, analyses them into elements whose value seems accidental and of no weight. Man, the beholder and listener, has been dissolved into two parts or two aspects, a mind which can find no *reason* to give weighty value to the contemplation of these things as an end, and a machine which *qua* machine is not occupied in evaluation. And the physical in all its order and wonder has been regarded as never in any respect or element having value in itself.

I have begun by considering some of those pleasures most vaunted as noble, in which man is at his most contemplative. But now let us consider man in his activity. Strange would it be to vaunt the pleasure of listening to music, or of watching the ballet, and not also to set value on the playing of instruments or upon agility, grace, strength and vigour in dancing, ice—skating and the like, pleasures in which there is realised a certain unity of man in his body with his musical instrument, as the rider with his horse, a unity found also by the craftsman in his craft. When the activity involved is physically specialised, as in the playing of musical instruments, no moral worries are raised. But what if the whole body becomes involved, hearing blending with rhythmic movement of the whole body? Then we become worried: criteria are needed for distinguishing when this involvement becomes analogous to drunkenness, or when it carries with it the danger of such mob hysteria, as we can still sense when seeing films of Hitler's Nuremberg rallies. But such bodily involvement is not inappropriate to us. We need to apportion our time rationally, but it is not bad to take pleasure playing in the sun, or (less sleepily) actively to bring one's whole body to a feeling of unity with the sea and the waves, a mixture of *Yin* going with the waves, and *Yang* exercising mastery amidst them (to use the Taoist words for two aspects of human living and action), and a skier, e.g., almost makes his skis parts of himself, as it were, experiencing them in unconsciousness of separateness in the activity of ski—ing.

I have spoken here not of creative productive activity but of what sounds like recreation: but some recreation is closer to contemplation than we suspect, embodying *appreciative knowledge* in a bodily aspect. But all this depends on a sense of unity in us, upon our not thinking of our bodily activity technically, upon not abstracting body from mind.

Ease and peace in unimpeded bodily activity — such as are found in the able—bodied in walking and other uncontemplative forms of relaxation, as in stretching their limbs, lying on the grass, and suchlike — are probably also found in their natural expressions of emotion, whether love or anger, and are largely lost to some of the disabled. It is worthy of note that this pleasure in psychosomatic unity and well—being is much more important to happiness than is freedom from pain or from spasticity. The danger of oscillation between frustration or despair on the one hand and successful achievement resulting from an almost manic exercise of control is greater for the disabled in the measure of the impairment of their capacity for relaxed physical existence and activity, part of the naturally given normal key to serenity and peace. I would mention this further to illustrate the psychosomatic unity which constitutes the background and basis for the activity of craftsman or artist.

You will notice that my discussion of contemplation and appreciative knowledge has brought me to the conception of a bodily activity, such as dancing, swimming, playing a musical instrument, which might have intrinsic value — and even in the fashioning of a work of art or craft, the fashioning as well as the product may have intrinsic value for man. And we are not separating pleasure as an end from the activity as mere means, external to each other. The satisfaction or pleasure is in the activity, inseparable from it: the mode of the knowledge which is enjoyed or which satisfies is that mode of knowledge which the intender has in his intentional activity as such [7], as internal to it as knowledge of seeing is to seeing; not external as in the relation of see—er to the object of sight.

This is the value, not of the useful, as such, nor of the pleasing as such, although liable to give rise to both of these, the usefulness being dependent on practical conditions and the pleasingness arising logically in virtue of the logical connection between the intrinsic good or *honestum* and pleasure if (e.g.) psychological disorder, injury or death do not interfere with or prevent it.

Section 5 : The Structure and Intrinsic Value of Human Living

I have taken as my starting—point the consideration of enjoyment and the contemplative aspect of man's nature because it seemed to me the proper way of introducing this notion of an activity which should be considered to have value in itself, intrinsically. (I have not considered aesthetics in that sense in which it is occupied with art, partly because in order to understand art one has to understand that playing with the brush or expressing human emotions, even when they do not give contemplative pleasure — though they do give appreciation — are also examples, quite probably, of such intrinsically valuable activities. So it is evident that in art we have the aim, not only of producing objects that are beautiful, but also objects that enable others to appreciate human emotions; and we may have a variety of other purposes. It therefore seemed premature to begin by considering aesthetics in its most general aspect, which includes the theory of the value of works of art, since this would bring extra conplications into our discussion, not relevant at this stage to our main line of inquiry.)

There are a great variety of human activities which cannot be properly understood except if it be conceded that they have some value intrinsic to them. It might be that eating serves practical nutritive and also social functions, and maybe that apart from the openness to the nutritive aspect a pleasure in eating is in some way false. (One has in mind the custom of the Romans who vomited in order to eat more afterwards: this seems to represent finding a false pleasure in eating.) Nonetheless, eating for pleasure, when not thus corrupt, seems perfectly legitimate. And if we consider the act of heterosexual intercouse this likewise appears to be an act with intrinsic value in itself. However, because of its connection with personal relations, I shall reserve more particular discussion of sexual relations until later except at this stage, making the observation that it can be conceived of as including a contemplative aspect, brought out in the Anglican Prayer Book in the marriage service where the words occur "with my body I thee worship". This brings some analogy with man's love of nature, a swimmer's love of the water, the walker's love of the land and the sky, on the one hand, and, if God exists, love of God as well as analogy with loves of other human beings, into view.

I mentioned four key areas of human fulfilment: appreciative knowledge, creative activity, personal relationship, and play. I do not mean to consider play separately since it seems to me to arise everywhere wherever enjoyment arises and is the fruit of expression of enjoyment and appreciative knowledge. Always it has to be remembered that for true human happiness these four things need to be integrated in some intimate way.

If a fully critical account is to be given of the notion of a human activity which has intrinsic value, i.e. an account which deals with the most common and obvious objection to this notion, it will be found to highlight the character of human living as an integrated whole.

This most common and obvious objection of the philosophers runs as follows: "What is the good of it?" in regard to any of these activities which I have named as intrinsically good. One might suppose that some individual intuition was the only justification which could be offered in each case. However, this neither seems all that is possible for us to say in response, nor does it mention the structure of human life wherein these intrinsically good activities are integrated together. My view is that one should say rather that there is intrinsic good in human living as such, which is one species and mode of animal living. In Aristotelian terms, the *ergon*, work or function of the human being is living humanly, not in the sense of being alive rather than dead, or of being of the class of animate rather than inanimate things, but in the sense of exercising the activities of life, in particular the activities appropriate to human life. And, within this activity of living, these elements have intrinsic value as appropriate or proper parts.

I envisage no divorce between the notion of intrinsic good and the concept of desire because I regard the notion of good as tied to the notion of enjoyment, as the appropriate object of enjoyment, i.e. the object of enjoyment when disorders or defects of various identifiable kinds do not enter in. Obviously there are elements in human life which are integral or necessary and pointful without being in a normal case, except accidentally, pleasurable. Defecating has been mentioned as a possible example. It seems essential to any intrinsically good activity, considered individually, that it be both pointful (as eating has nutrition as its point) and typically pleasurable.

Judgment as to whether an activity is a proper part of or quasi—integral to human life (if not necessary in the case of every individual) or whether it is an extra, accidental to this core, like making wheels and ships, is not always easy. Reading and writing seem to me the most difficult middle sorts of case. Often the situation is that one has a general kind or genus of activity (a determinable) which, as such, is quasi—integral to a human living (e.g. communicating, making a joyful noise, using tools or instruments, exploring, entering into activities jointly with others for the sake of pleasure...) which have to be realised in some particular ways, but none of whose particular kinds of realisation are necessary. In these cases the more specific kind of activity (the determinate), e.g. writing letters or books, singing or playing musical instruments, dancing, or playing a particular kind of game, will be intrinsically good, but not quasi—integral or necessary.

The principle of structuring which influences one in judging which activities individually are intrinsically good simply in virtue of being proper parts of human living, and so as the primary intrinsic good in their areas, will doubtless also influence one's view of what counts as living well or the way the fortunate man or woman lives, i.e. one's view of happiness.

The significant point to notice in this whole approach is that I am taking the *ergon* of man to be human living as a modality of animal living, typically involving the body in an integral way, rather than solely in terms of *nous*, if this be considered in purely intellectual terms.

What we need to note as a consequence of this is that it will be characteristically in one and the same move or act that human beings will violate or realise their animalhood, their humanity and their personhood.

The effect of this approach will be that in regard to intrinsically good activities (although some general exploration of their place in the general structure of human life, well—being and happiness may be offered in response to the question "What is the point of it?") there is no general need for a more particularised, means—to—end or technical answer, and often the provision of such an answer may even imply some defect in a person's mode of living or attitudes — as when a person feels that he has to offer some excuse for enjoying a view, enjoying eating, enjoying making love to his wife... On the other hand where activities deviate from the norms of what is intrinsically valuable, their motivation (it seems) has to be technical and there may arise questions as to whether, as a result, these may not, by their presence in a person's life, be symptomatic of some defect in the mode of life concerned. Thus, the person who uses drugs or undergoes surgery in order to perform better or for a longer period in sport or singing may have violated some deep norms in human life and in one act violated their animalhood, their personhood and their humanity, and, along with these, their communal nature.

Section 6 : The Strangeness of a Purely Technical Approach to our Basic Activities

We are now in a position to observe the strangeness of any purely technical approach to these intrinsically good activities in which human living finds experience and fulfilment, especially when these activities are "basic" in that sense in which thinking, speaking and communicating, seeing and hearing, walking and exploring, using the hands and eating, making love and having children are "basic", while dancing, ski—ing, and mountaineering are not. In order to illustrate the strangeness which I have in mind, it will be useful to consider an example which is at present completely odd and out of the way, an example which is therefore not currently the subject of discussion or controversy. This example serves to illustrate the problems with which we may have to grapple today only in academic—seeming philosophy, but which (to judge from parallel cases in the past) may in some unexpected way confront us as men and women.

Some wild debater might suggest in respect of eating that population problems might force us all to a diet largely consisting of a soya—bean derived paste, the consumption of which by mouth had no advantages, whether social or to palate. He might suggest further that then we might come to prefer to obtain nutrition by intra—venous injection of some yet further processed derivative product. He might argue that intra—venous feeding was permissible when normal oral feeding was impossible, and that in special cases it was clearly a life—long necessity for some people owing to defects or loss of gullet or stomach, in order to soften up his readers or hearers to acceptance of his argument.

Now you will note that we are being asked to consider numerous quite extraordinary and implausible hypotheses, and that we would readily object to the argument from conceivability to real possibility, or from real mundane possibility to possibility within God's Providence. So, this whole approach in morals will be subject both to metaphysical objections (as well as physical, biological, psychological and sociological ones) and to objections deriving from a possible bearing of theology upon morals.

However, it is not clear that we should allow our debater to be confronted only with these methodological objections to argument in morals from mere hypothesis, for if we thus confine ourselves, then we will have allowed him to get away with the suggestion that the reasons for preferring oral feeding to intra—venous feeding are reasons solely of pleasure and convenience. He may for instance have been willing to discuss the unhygienic character of mass injections, the need to take measures to prevent damage to unused digestive tracts, and to dampen hunger sensations due to emptiness of stomach, the loss of pleasures of tongue and palate and the loss of the social pleasures surrounding eating. I.e. we would have allowed him to treat the question of the preference for normal feeding over intra−venous procedures as a merely technical one. The question needs to be asked whether this response is right and, if not, why not − on the fact of it to respond in this way is to agree to conduct the dispute with our wild debater entirely on his own terms and as it were to give the game away at the start as to how the condition of human life on the whole is to be viewed, i.e. whether it is to be viewed purely technically or not.

Or, again, imagine that for some athletic or industrial purpose it were of advantage to have very long arms or extraordinary shoulder mobility, or for some other purpose a very long tongue. And imagine that the desired effects could be achived either by a programme of exercise and diet over long periods, perhaps from early childhood, or by surgical operation, ingrafting extra tissue or tissue substitute at relevant points, or by drug use over shorter periods. Is the sole objection to the latter two types of procedure the tendency of such interventionist approaches to have unforeseen side−effects, nature being, as it were, wiser than man? Or is there some distinction between approaches which restore nature, or which, using it, assist it and improve it, to ones which supposedly improve it only by violating it in some way? Are some procedures just wrong not solely because they are unsporting or dangerous, but because they are unnatural? Is castration to achieve certain results in singing not merely a gross offence in justice against minors, but also wrong because "unnatural"?

The whole area I have been viewing is problematic.

Consider the following bodily activities which may be thought to be not inappropriately valued, *qua* activities, and not only for the sake of consequences aimed at: dancing, eating and copulating.

Now, we see no problem about dancing for almost any legitimate end, even the getting of money. In the case of eating, we find making oneself sick in order to be able to eat more has more than a sniff of perversion about it, unless perhaps the eating more is necessitated by some important end: e.g. one is an ambassador in peace negotiations, and does not wish to appear to reject the forms of hospitality customary amongst one's hosts. To go in for this practice, on the other hand, in a fair for the sake of money gained by entertaining others would seem rather corrupt (and also possibly symptomatic or causative of corruption in the spectators). But in the case of copulation, we seem to prefer irregular extra−marital intercouse between adults, even spontaneous or promiscuous rather than high−minded, to prostitution; and the objection that the ponces, not the prostitutes, get most of the money is entirely subsidiary.

In short, we are very reserved about *any* merely technical motivation in copulation, far more than in the case of eating: but even in the case of eating we have deep reserves about technical uses in which the nutritive aspect is not respected at all − we don't mind the eating of unnecessarily large amounts for pleasure, if the food is still geared to possible nutrition, but find the Roman practice of eating for pleasure or society with the intention of preventing nutrition by making oneself vomit base and degraded.

Again we are more worried about the substitution of artificial insemination for copulation than about intra−venous feeding for normal feeding when it is suggested that these are the only means of obtaining the ends respectively of children or nutrition in particular cases; but when there is no such suggestion of necessity both seem disordered as human activities.

The differences between our attitudes to copulation and to eating are not, I think, *simply* due to the gravity of the matter involved — why so grave, anyway, if the possibility of children does not come in? If a generative meaning is intrinsic to copulation, this would be relevant to the consideration of gravity.

In any case copulation, unlike eating, is *intrinsically* other—person involving: firstly a partner, and secondly possible children. This other—person orientation has provided some philosophers and novelist with a further way of thinking about sexual relations: they have envisaged the possibility of telling lies with one's body, as if by one's biological nature one could *mean* one thing by an act, even though by one's controlling mind (thinking of the body as an instrument) one pretended, in an act of "bad faith", to mean something else. But how can this conception, that, by the very biology of his intentional act, a person *qua person* means something determined by that biological structure, be justified?

We find that many people find it obvious that sexual relations should be "fully personal". Let us agree wholeheartedly! But why this requirement on sexual relations (so as to imply that they are *wrong* in other contexts) when we do not place such a requirement on many other relationships? And what is this being "fully personal" going to involve?

At this stage I have set forth questions, not offered answers — but clearly on any view the question will arise, what intrusions into the area of sexuality and generation are to be rejected on the ground that they embody some rejection of one's animality in a pretence that one's sexuality can in this or that respect be ignored or truncated in some aspects, i.e. on the ground that they deprive man of *honestas* in respect of his animality? For the appropriate way for human beings to live as animals is not to proceed by imitation of this or that other animal, but to proceed after reflection in a way which also accords with what belongs to their animalhood and with acceptance of this form of animalhood as of value in itself.

In the discussion we have just been following through, I have in fact been giving a background for an introduction of the concept of the unnatural in human acts. In modern times, as I pointed out at the beginning of this paper, moral philosophy has been conducted as if all that matters in human morality arises either from some intrinsic value which ends or actions would have for any intellectual being whatsoever, non—bodily beings such as angels equally with men, or from the utility of things as means to the getting of these intrinsically good things. Thus there is no argument in Mill to justify his view of the objects of moral obligation as including all human beings independently of the stage of the development of their faculties. And Kant's moral philosophy is geared at its most basic level to cover every rational being as such, so that those influenced by him have dwelt upon cases of contract—breaking, lying and such—like, and treated killing and suicide as if they involved the termination of existence as such, even as mere souls, rather than as the termination of existence or life in some biologically defined sense.

Morality of this kind, as I pointed out, is ill—equipped to explain the point of various burial customs, the significance of death, or even of pain, the obligations of kinship or the identity of the objects of moral obligation.

What I have attempted to show is that these defects are connected with the failure to recognize the significance of man's animalhood. From the standpoint of a holistic view of man such as I was outlining in Sections 1 and 2, it amounts to a kind of *reductio ad absurdum* of any moral theory in regard to the conduct of human life if it includes only general principles supposedly applicable to all intellectual beings as such together with technical consideration of the adaptation of means to ends.

Section 7 : Is the Concept of the "Unnatural" Relevant to Medical Ethics? Consideration of Examples

In ethics, we often achieve a greater realism by keeping concrete examples in view. In medical ethics, many questions are raised which immediately and directly concern the value of life as such: questions concerning euthanasia, whether of the old, those impaired in faculties, or those grossly disabled in war or accident; abortion, and the variety of grounds on which it is proposed; war insofar as it has medical aspects and — under the heading of contingency planning — consent to nuclear warfare, I am not concerned with these in this paper. I am only concerned with questions of the manner and style of human living, not its termination.

The examples which confront us are doubtless almost without number. I mean to pick out for mention only eight. These eight are:

(1)   sport;
(2)   the interests of public projects, whether in entertainment or in industry;
(3)   the use of biological and chemical techniques in treating deviants and dissidents;
(4)   the extension of cosmetic surgery;
(5)   the use or abuse of drugs (for 'kicks' or other purposes);
(6)   the co-operative use of medical techniques in masochism or masocho-sadism;
(7)   the substitution of insemination for copulation, or any such fragmentation of the sexual act;
(8)   the use of biological and chemical techniques in warfare.

In the life of most ancient civilisations — and indeed almost all civilisations apart from modern Western so-called materialism — civilisation did not mean or involve an attitude of mastery towards nature, environment and the body, but allowed a certain empathy and willing co-operation and respect for the rhythms and structure of nature. This is expressed in the incorporation of such respect and empathy in ancient rituals and mythology, in the role of gods of fertility and of the household in Roman, Celtic and Norse religion and goddesses of the same in Greek religion. It was also expressed by the importance set upon Wisdom or the Torah in Semitic tradition. Thus, even where in Indo-European and Semitic thinking the male principle held a predominance over the female, this did not signify or involve the arbitrary dominance over nature and the body, conceived of as female, by mind or spirit, conceived of as male — as it were, to borrow Chinese terminology, the exaltation of *Yang* at the expense of the suppression of *Yin*.

Only modern man, in the wake of a certain individualism, certain conceptions of law, mastery and control, and a certain success in technology and in the financial and social organisation of resources, has felt able to suppose that anything is possible for man (e.g. that he will always by some means or other be able to find new sources of energy, contain the problems of pollution, etc.), and that, granted that some end is socially good, not only will its achievement be possible, but also there will be no limits set from the side of nature or the bodily on what it is possible to do in pursuit of this supposed good.

To consider the types of example I mentioned in order, let us consider first the area of sport. It is evidently permissible for laboratories in different countries to enter into a certain rivalry in prestige projects of certain kinds, e.g. in the production of cars, speed—boats and such—like, intended to achieve certain speed records or win certain races, and to design vessels to take men to the moon or planets, provided that human resources are not thereby misused, and basic goods are respected. But the conception of sport gives a key role to the individual sportsman or sportswoman or team. The moment the use of drugs or surgery in non—restorative roles enters in, the character of the competition ceases to fall under the heading of sport, and the athletes of different nations become prey to pressure to co—operate in the abuse of their bodies in the pursuit of national prestige. They are at present protected by the conventions of international sports bodies. If they were not thus protected it would be a question of ethics as to what it was legitimate for medical or paramedical persons to do, or co—operate in, in the use of drugs and surgery in sport.

The view that in this area whatever is acceptable to society is acceptable humanly is broken—backed. This view offers no clue as to what should guide "society" in resolving such questions. It leaves it totally obscure how "society" in the relevant sense is to be identified, no reason why one should not rather be guided by the "better and saner part of society", and no reason why the individual's good or their preference should be guided by or over—ridden by that of society.

I mention the area of sport first because, although some cases (e.g. motor racing) carry with them a complex involvement with technology, it does nonetheless set in stark relief the question of what the aims and means permissible to medical and paramedical persons are to advise a person on diet, or programmes of training, or to attempt to restore muscles or skeleton to function, and such—like, either under the heading of restoring nature or of improving nature by means in accord with nature. Certain uses of drugs and surgery fall plainly outside this ambit.

When it comes to other public projects one might consider that in the area of entertainment (or even worship) the classic example of a violation of nature lies in the case of castration for singing purposes (a practice that lasted in Italy well into the nineteenth century). True, this represented a violation of the rights of minors; but, quite independently of any question of age, castration for the sake of singing represents a mutilation of a man in respect of one of his essential faculties for the sake of something not thus integral to humanity.

And one can imagine any number of motivations in industry (including space projects and such—like) for surgical adaptation of the human frame, and uses of drugs, aimed not to maintain, but to subvert normality. In every such case the ethics of some medical and para—medical procedure will come into question.

These cases we have considered present themselves, at least on the face of it, as cases in which, in the interests of some evidently subordinate human aim, whether success in so—called sport, achievement in entertainment or in an industrial project, some violation of the environment or of human nature, especially in its biological aspect, is involved. But parallel questions arise when chemical and biological techniques (counting, for instance, techniques involving interference with sleep or concentration as biological) are used to 'treat' deviants and political dissidents. In these cases the goal is not some limited subordinate end, but related to the coherence of "society" itself. Here above all, in this initial group of types of example, it is vital that the medical or para—medical person limit his procedures to ones respecting the personhood and body of the supposed deviant, and limited to restoring him or her to proper human capacity to function, and not aimed at more general goals of the society in which he or she lives. It is not, and must not be, regarded as the function of medical or para—medical personnel to bring the persons committed to them into conformity with the prevailing ideology. Nor is it legitimate for them *or for anyone else* to act in a way which does not respect their personhood and their bodies, i.e. in any way which damages in any respect those committed to their care.

I have spoken first of types of example involving public interest or public authority. But in the area of private or privately directed medicine, cases arise in which individuals (or pairs or groups of individuals) seek medical or para—medical assistance in projects which are not publically directed, firstly in various extensions of 'cosmetic' surgery, secondly in the use, or perhaps abuse, of drugs whether for 'kicks', or for 'tranquillization', or for any other purpose (we have mentioned sport), and thirdly the co—operative use of quasi—medical techniques in masochism and in other practices traditionally regarded as perverse.

At least in the first two cases a considerable range of distinctions will need to be made. The elimination of Mongoloid eyelids, which might be objectionable if expressive of racialist prejudice, might serve a legitimate purpose in the case of disguise and escape, and then be legitimate. It does not seem to represent, for instance, a repudiation of one's human animality in the way that an operation to change the sex of a person whose sex is already determinate would do. (Of course, if what is called a change of sex operation involves only, in some particular case, the making more determinate of what is biologically indeterminate, then this would not come into our present discussion.) To be determinately of one sex, rather than another, is an integral part of the human form of animality. As to the use and abuse of drugs, the case is yet more complex. We need first to remove some of the areas in which vagueness or obscurity arise.

It seems evidently legitimate to utilise various processed materials, spices in cooked foods and wines, for instance, to increase the pleasures of eating and company; besides that, a wide range of medical purposes have long been served by products not formally classified as medicines. Yet if the drug concerned causes damage or induces a state in any degree settled of false consciousness, depression, inactivity and withdrawal from communication, or is taken for the sake of sensations, tranquillity, athletic, academic or other work—related purpose regardless of such consequences as these, it seems illegitimate.

I deliberately avoid picking out some supposed hit list of drugs never to be used (here someone might have proposed lysergic acid, heroin and cocaine, to mention the most obvious seeming) unless it be in some strictly delimited medical context, while leaving some others (e.g. caffeine, alcoholic drinks, marijuana, monosodium glutamate, amongst the most common candidates for social approval) exempt from criticisms of the type with which we are now concerned. Such an approach leaves the definition of proper 'medical' use quite obscure and suggests that the evils associated with caffeine, alcohol, etc. are only those of excess, restricted lifestyle, misuse of time and money at the expense of one's dependents, so that, for instance drunkenness (whether on particular occasions or habitual) is to be viewed in the same light as gluttony — of course everyone is aware of the problem of addiction and controversies concerning its nature, but on almost every view this is not intrinsic biologically or psychologically to the regular overuse of alcohol or to drunkenness, even when habitual. Yet, such an approach seems misguided: drunkenness itself seems to represent an abuse of the human psychosomatic or animal system of a more generalised kind than overeating.

What we need is a definition or account which allows us to discern more precisely when a certain use of some particular drug or drugs represents an abuse of the body or, to use different words, an abuse of the psychosomatic whole which a human being constitutes, and when it does not.

In some cases there will be some legitimate or even laudable end for the sake of which the use of the drugs under consideration is proposed, but in which the fact remains that the use of drugs proposed involves such settled damage or injury to faculties, or such settled distortion of psychological state or attitude, as to be excluded, however laudable the end. The principle of totality, according to which in the interests of the overall functioning and good of a person a faculty or even a life may be sacrificed, as when, because of a cancer, organs vital to some faculty, or an organ containing a foetus not yet viable on its own, are removed, must not be misconstrued or interpreted in a way that makes it equivalent to a utilitarian approach (that is an approach in which the rightness and wrongness of actions are judged according to the consequences to be anticipated from them). Man does not have such a command or control either of nature or of his fellow human beings as to make it right for him to arrogate to himself responsibility for each element or aspect of the predictable, likely, natural or actually anticipated issues or 'consequences' of his actions. He is not God to carry such responsibility; and subsidiarily (it may be remarked) lacks reliable knowledge of many of the most weighty consequences (some distant and some close) of his actions. Therefore, the principle of totality, so—called, properly used does not invite the taking of an overall view of the whole of that person's future life in its 'quality', or such global view of the life of the family or of the community of which the person is a part. Rather, it has proper application only when the dilemma confronting the human being presents itself in some way of which it is absolutely clear precisely that it is not thus grandiose or global in its conception and planning, but has as its subject a question in certain definite ways quite limited in scope: the action proposed, under the first description under which it is elected upon and chosen, has some limited character, e.g. the removal of this cancerous organ, and it is in virtue of this description and not in virtue of some more extended situational account that it is identifiable as directed to the overall good of the subject(s) of the action.

These will be cases in which the means required render the action wrong, even though the end is legitimate. In other cases, it is the end (that is, the thing set upon, inspiring action, which is not just chosen but desired) which is wrong or, as we commonly say, perverse. Thus, if the end which the heart is set upon is oblivion, or the confusion of faculties, or withdrawal from communication, or a settled state of false consciousness, or of withdrawal from choice or activity, or excitement or semblances of rapture unrelated to any appropriate object, 'kicks' or sensations which are not from anything worth getting a 'kick' from, then the end is wrong or perverse.

My first three types of example concerned public projects and the techniques permissible to attain their achievement. My second three types of example concerned private projects and the techniques permissible in the pursuit of pleasure. My seventh example is not concerned with the use of techniques to attain pleasure, but rather the use of techniques to attain the end of copulation, viz. the having of a child by a mother by means (viz. artificial insemination or *in vitro* fertilisation) which are precisely of such a nature as to exclude the pleasure commonly enjoyed in copulation. It involves the substitution of two technically motivated and medically guided or contrived pieces of performance by man and a woman for a single co—operative psychosomatic act of the sort we represented as intrinsically good and naturally pleasurable. For a child to originate in such an act seems to be an element in the type of family bond most to be desired. And it is something which makes the family a characteristic expression of the human form of animality, and, one may equally well say, of the integrated psychosomatic (ie. precisely 'animal') character of humanity, rather than analogous to such social arrangements as business partnerships and political organisations, not thus founded in a psychosomatic act on the part of the founding members.

Wherever human beings suffer some privation or are frustrated from attaining some normally obtainable end, the question arises as to whether the means proposed to overcome the obstacle to attaining the end concerned count as a proper assistance of nature, restoring or extending the powers of nature by a proper use of nature, or whether they represent a certain *hubris* or arrogance, the expression of a domineering attitude to nature, in which the human being is involved in an implicit rejection of his human condition and the limitations set by being an animal, set by the respect of *pietas*, due to his fellow human beings, his own body, and his whole environment (including, of course, his fellow living beings). It has been held, and is still held by many, that to separate the procreation of children from the sexual act embodies in itself a violation of the human form of animality: a violation in which, in a domineering way for the sake of securing the end, the having of children, the sexual act in its normal relation to procreation is set aside, or not used because it cannot be used. Instead, an alternative way of producing children is seized upon, a way that

—26—

leaves them without their properly human psychosomatic context or origin and makes them the products of technology, a technology in which many parties have an intrinsic role.

It is plain that this view is much disputed; but it is equally plain that it is only the setting aside of this view or of this insight that has opened mankind to the spectre of the development of a kind of technological and eugenic handling of children, and indeed human beings, as products, of which Aldous Huxley's *Brave New World* gave some dim foresight. Perhaps, if modern man had not come so to conceive his lordship over the world as giving him the freedom of a master or despot (rather than involving the care of a governor for the community he governs), and if he had not therefore handled his environment, the community of living beings and the balance of nature, with such absence of respect, he would not now find his perception so dulled in considering the handling of his own life.

I mention this case, not because the absolute principle I mentioned will be acceptable to all, but because it should be plain to any reflective human being that, even if the line of discrimination between what is in accord with and what violates our humanity and our animality is to be drawn differently, there must nonetheless be some moral principles affecting and sometimes restricting human action. And in particular there must be principles in regard to the procreation or production of children which arise from man's psychosomatic and animal character as a member of biological nature, and not solely from his character as an intellectual being.

Lastly, we need to consider one particular kind of meaning which is given to the words "clean" and "dirty" as applied to methods of warfare. There are of course meanings of these words which refer to the extent to which innocent life, the lives of women and children, and of hostages and prisoners of war, is disregarded, and also meanings which refer to the extent to which the methods adopted are in some peculiar way, not conventionally reckoned with in traditional warfare, deceitful or underhand (perhaps certain ways of making use of the entry to the enemy territory gained by an embassy of peace, or the opportunities provided by a truce or by some apparent offer of mercy). But there is another way in which these words are used: nuclear bombs which bring about a more noticeable degree of genetic and slowly maturing disease, damage and mutilation than other bombs of similar explosive power are described as "dirty", and regarded as peculiarly objectionable, in the beginnings of the way in which biological and chemical warfare are regarded as objectionable; to speak metaphorically, they do not just destroy and kill but get "into and under the skin", they contaminate and corrupt nature, that is the environment and the body, as it were to bring to realisation a type of 'living death' in which starvation and other disasters resulting from the social and economic chaos brought about by war could become dwarfed by the phenomena of disease and deformity.

It may be that the motivation of the great powers in (it is said) not entering into the fields of contingency planning of chemical and biological warfare, in the same way that they have entered into such planning for nuclear warfare, stems from the more evidently certain uncontrollability of their effects on the user. But it remains true that the horror, the abhorrence felt by almost every man for these things stems from some deeper source, namely the sense of the deeper intimacy of the disordering and injury involved in these attacks on man's biological nature (to be conceived of as a psychosomatic whole) than any injury or disorder which man is able to envisage as having the character of something external to him — as it were something in relation to which he has an independent standing for purposes of deliberation, e.g. as to how it is to be coped with, as an external attacker is coped with. It is not only those intrusions into man's life which, as in brain—washing, are felt by man as attacks on his "inner self", but also things of this biological character. They do not thus aim at the subversion of the mind, but nonetheless surely at securing a psychosomatic effect, in giving the human being reason to be disgusted with himself and with the human, and weak in the resources to be other than weary with life.

I have brought into view examples of these eight different types in order to make it plain how implausible it is to suppose that the concept of the "unnatural" has no place in ethics.

By "unnatural", in the use of the word here relevant, is not meant the unusual or that which is not normal, nor the 'artificial'. There are many special circumstances which require unusual actions and it is quite typical of the good to be fresh in a way which makes it, in the particular case, unusual and 'not normal'. And it is a proper part of human nature to use tools and instruments in work or play, and technical devices to achieve his ends, including the ends characteristic of medicine.

Nor does the word "unnatural" here mean 'what is contrary to natural law' since this signifies nothing more than 'is in principle naturally knowable to be morally wrong'. This includes many things, murder, promiscuous or adulterous heterosexual intercourse, lying, betrayal of friends, breaches of contract, breaches of trust in general, dispossession of others, of the food, means of livelihood, etc., which they have proper use of, etc.

Rather, by the word "unnatural" we mean that which violates man's psychosomatic, or biological, nature, i.e. that which, *qua* means, or *qua* end runs contrary to his character or standing as an animal, with all that this involves, both in the functioning of his own life and of his relationships to other human beings, and in his regard for and treatment of the rest of nature of which he is a bodily member. A morality for man which pretends to be a morality applicable indifferently to every personal being as such, or to every 'rational being as such' (a sort of morality for angels and devils) is an excrescence of modern times. In this excrescence the material and physical are conceived of as in themselves lifeless and purposeless, so that it is left to mind alone, free and untrammelled from below, to arrange and direct them as it sees fit — a conception which is the offspring of western dualism and individualism and of Roman imperial conceptions of lordship or mastery. This conception has removed all inhibition, reserve or qualification from the expression of a purely technological and domineering attitude to the natural and the natural world.

In the era in which we live, the respect paid to the values and interests of the "community" and to publicly approved subjects, the rights accorded to private desires, the development of a quasi—capitalist (product—orientated and production—planned) attitude to the shape of special groups (including the family) and to the procreation of children, and finally the emergence of the conception of total war, have made it certain that, in the context of the development of medicine and psychiatry, the presentation of the properly *human* character of human life will turn on the acceptance of values and of limits to action set by man's psychosomatic and animal nature. For man, a morality devised as if we were angels or essentially just minds, selves, or persons is a morality calculated to bring mankind stage by stage into a state best described as a 'living death'.

Notes

[1] The best introduction to this notion of happiness as a structured whole remains Aristotle's *Nicomachean Ethics*, Book I, supplemented by Book X. An acute criticism of empiricist conceptions of happiness is to be found in R. Simpson, "Happiness", *Am. Phil. Q.*, 1975, drawing out the Aristotelian picture against rival atomistic modern conceptions. This picture of happiness as a structured whole is developed in my Section 5 below.

[2] The isomorphism of mind/brain identity theories and dualism is highlighted by Ryle and noted by Geach. In my exposition below of the rival holistic perspective, my exposition is mainly shaped by Ryle and by Aristotle. But this exposition proceeds against the background of the more precise arguments of Strawson in the first part of *Individuals* and, earlier, of Aquinas. The consequences for ethics have, in recent times, been emphasised in papers of J.M. Cameron, and popularised in the books of Fritjof Capra, as well as in many other expositions, but without adequate precision in respect of the wholes which are not just an aggregate of parts (persons, states, or the human race) or else in regard to the distinction between metaphysics and ethics.

[3] In Aquinas' technical use of the word, the object of "love", since his use of the word "love" is extraordinarily general and not tied either to *eros* or to loving—kindness (Hebrew *hesed*, Greek *eleos*, Latin *misericordia*).

[4] Indeed modern moral philsophy has even seen it proposed that the meaning of the word "good" is to be explained in terms of the notion of commendation, commendation being understood as some sort of recommendation in respect of action, so that statements about what is good and bad are to be understood by analogy with imperatives or prohibitions. Since to commend or to praise something seems to be precisely to explain some way in which it is good or to state that it is good or excellent, it seems almost perverse to attempt explanation in the opposite order. Besides, it is only affirmative ascriptions of goodness that would be straightforwardly explicable in this way, even if this mode of explanation worked at all, whereas more complex occurrences of good in sentences, e.g. in the antecedents or conditionals, cannot be so explained, as Geach has pointed out in "Ascriptivism" (*Phil. Rev.* 1960, Vol. LXIX, pp. 221—5)

Of course the background of this view is the idea that the meaning of an individual word, prescinded from any context of statement or sentence, can be explained in terms of the role it plays in human life (a Wittgensteinian derived point of view); but if we are to consider and understand human action as a whole and relate it correctly to conceptions of desire and intention, it is, I believe, to be argued that the concept of good has to be taken as primitive and used in order to understand all these other concepts rather than vice versa, just as in the case of propositional knowledge we have to presuppose the concept of truth.

[5] *Man at Play*, Hugo Rahner (Herder and Herder, N.Y., 1972).

[6] *Beast and Man: the Roots of Human Nature*, Mary Midgley (The Harvest Press, Hassocks, Sussex, 1979).

[7] The fundamental insight in respect of the unity between the pleasure and the activity in which it consists is Aristotle's (*Nicomachean Ethics*, Book X). The same thesis is upheld by Ryle and Kenny, but without resolving the difficulties in it. I have attempted briefly to indicate how certain insights of Anscombe in *Intention* and of phenomenologists such as Merleau—Ponty may be helpful in resolving these difficulties.

3 Objectivity and alienation: towards a hermeneutic of science and technology

SIMON GLYNN

I Introduction

The single most striking fact of the modern age is surely the ever increasing disparity between the possibilities that confront humanity and the level of our actual achievements. In an age in which we have more knowledge and a greater technical capability than ever before, starvation, resource depletion, pollution and thermo−nuclear weaponry threaten the survival of the population and even the planet itself as never before. It would, however, be quite wrong to conclude that this state of affairs is the result of some oversight, inefficiency, maladministration, or even plain malevolence. Indeed it appears that our inability to close this gap between the actual achievements of our technology and its potential increases almost in direct proportion to our knowledge, irrespective of the motives of those who utilise this knowledge.

Phenomenological and existential philosophers have not been reluctant to address this problem. The explication of the roots of this crisis constituted the final phase in Husserl's works; and the interpretation of its significance was a central theme of the existential phenomenology of Heidegger. Both in different ways, see the root of the problem as lying in the modern world−view; and it is the nature of this world−view and its accompanying alienation that I now intend to explicate.

II The Alienation from Origins as the Origin of Alienation

Primitives, so the anthropologists assure us, find themselves born into a world steeped in mythical and metaphorical significance and meaning. Within this "life−world", the earth, the sky, rocks and rivers constitute an ancestral home resplendent with inherited and inherent meanings.

For modern, Post—Enlightenment, humanity, however, this unity between wo/man and world appears sundered. We confront a world apparently devoid of all *intrinsic* significance and meaning, a world in which all such *qualitative* elements have been eclipsed by the *quantitative* abstractions of Enlightenment philosophers and scientists. No longer regarded as "home", such a world, Heidegger tells us, is conceived instead as a picture devoid of any meaning and significance other than that which we see fit to attribute to it (Heidegger, "The Age of the World Picture", in Krell, 1978, and Heidegger, 1962, Section 44). Any such meaning and significance is consequently denigrated as "merely subjective".

As we shall explicate in detail presently, the same movement that apparently extricates us from the tyranny of tradition, from the demands and obligations of the ancestral "life—world", also seems to deprive us of our sense of belonging or "being at home" in the world. Apparently no longer *participants* in the world, we seem instead to have become *spectators*, whose knowledge of and relation to the world remains problematic.

Such "rootless" individuals, no longer regard the world as home or foundation of their existence, but see it rather as "*de trop*"; a world in which roots, the paradigm symbol of their belonging to, or being culturally grounded in, the life—world, now stand over against them, serving only to emphasise their own "insubstantiality" and to threaten them at every turn (Sartre, 1949, pp.172—3, and 1956, p.lxvi). From this perspective the apparent escape from the burden of tradition, from the ancestral life—world, is seen not as a liberation but as a loss of founding origins, a "casting out" from home. Thus the price of freedom appears to be our alienation, and we are, to use Sartre's phrase, "condemned to be free".

This primordial alienation from our origins, with its obvious theological analogue in the Eden myth, I shall call "Ontological Alienation". It is this ontological alienation which, as I shall show later, is at the heart of, is indeed ultimately nothing other than, the crisis to which I have already alluded. Before doing this, however, I shall first try to explore the way in which this alienation comes about; and where better to begin than at the beginning.

III   In the Beginning was the Myth

In the beginning, so we are told, God, having created Heaven and Earth, created Adam and Eve and placed them in the garden of Eden, where they were charged by God with "dressing" and "keeping" the garden, being only forbidden to eat of the "tree of knowledge of good and evil". However, in the attempt to "be as gods" they ignored this command, ate of the tree and "the eyes of them both were opened, and they knew that they *were* naked". Thus "afraid" before God, they "hid" themselves from Him. Seeing this, God knew they had eaten of the forbidden fruit and "drove out the man" from Eden, "lest he put forth his hand and take also of the tree of life and eat, and live for ever" (*Genesis*, Chs. 2—3). Let us now interpret this myth.

The first point to note is that it was not knowledge *per se* that was responsible for Adam and Eve's recognition of their own nakedness and their subsequent exclusion from the Garden of Eden, for quite clearly they needed practical know−how or "ready−to−hand" knowledge, as Heidegger calls it, in order to tend the garden. On the contrary, it is clear that "the tree of knowledge of good and evil" must have afforded them reflective knowledge, for it is only self−reflection that could have enabled them to perceive themselves at all; and indeed impure reflection at that, for as Sartre has pointed out, only in impure self−reflection does wo/man, the pre−eminent subject, mistakenly come to see her/himself as an object, as a reified thing (Sartre, 1956, pp.lii & 151−5).   Only in such impure reflection, therefore, could Adam and Eve come to perceive themselves as "naked" before the archetypal "look of the other", the gaze of their creating and sustaining God, who they consequently reconceive as a threat to their very existence, and before whom they are therefore afraid.

   In more general terms, then, as the myth clearly elucidates, it is in impure or reifactory reflection that the undoubted distinction between ourselves and our foundation is perceived as a real *separation*, with the result that this foundation is no longer seen to be such but is, on the contrary perceived as a threat to the apparently ungrounded self.

   However, while through such analytic or theoretical reflection, born of the forbidden fruit, Adam and Eve may indeed *distinguish* themselves from God, clearly He continued for some time to ground or sustain them in the garden, and therefore could not actually have been the threat He appeared to be. Indeed, the myth makes clear that it was not in fact the eating of the fruit *per se* but the attempt to overcome their *apparent* vulnerable nakedness by hiding themselves from God that *actually* led to their being driven out from the garden.

   By analogy, then, while by impure or reifying reflection we may indeed perceive ourselves as separate from our sustaining origins in the world and consequently come to see the world as Other, and ourselves as ungrounded and therefore vulnerable, nevertheless, while we are certainly distinguishable, we remain in reality *inseparable* from this world, which, after all, continues to be our very sustaining foundation. Clearly then it cannot be the world as such which poses the real threat to our existence. But, as in our analogy, it is, as we shall later see in detail, precisely the attempt to overcome a merely *apparent* threat posed by the world that is ultimately responsible, for our actual alienation, and may yet result in our damnation: "what has long since been threatening man...", Heidegger explains, "... is ... self assertion in everything" (Heidegger, 1971, p.116).

   Bearing the Eden myth or metaphor in mind let us now embark upon a more detailed exposition.

IV Ontogenesis and Alienation

Not unlike Adam and Eve, who were from the first subject to the prohibition of eating the fruit, men and women born into the life−world and engaged in practical activity routinely encounter recalcitrant natural facts and social taboos, characterised by Freud as the "Reality Principle", which serve to thwart the "Pleasure Principle", our attempt to enjoy the object of our desire. Such resistance results in reflection and the concomitant emergence, confirmed by Piaget's experimental results, of a reified concept of the self which, in misconceiving itself as radically *separate* from the natural and social world, comes mistakenly to perceive itself as vulnerable to those things and people which in reality, on the contrary, precisely create, define and sustain its identity.

V   Objectivism as a Subjective Accomplishment

The life—world as given in *immediate experience* is full of beautiful sunsets and threatening storms, lovable and hateful people, enjoyable and frightening experiences. It is, in other words, a world full of meaning and significance, in which quality and quantity are inextricably intertwined, a world quite unlike the so called "objective" world full of Newton's abstract mass points, or Faraday's force fields, and a world which is even more unlike that of Quantum Physics in which, according to some, waves representing mathematical probabilities are said to propagate themselves through a non—existent medium. Clearly then as Husserl points out, "The contrast between the subjectivity of the life—world and the 'objective', the 'true' world, lies in the fact that the latter is a theoretical—logical substruction, the substruction of something that is in principle not perceivable, in principle not experienceable in its own proper being, whereas the subjective, in the life—world, is distinguished in all respects precisely by its actually being experienceable" (Husserl, 1977, p.127). Increasingly, Husserl insists, we find the "... substitution of a methodically idealised achievement (i.e. the 'objective' dehumanised world) for what is given immediately as actuality presupposed in all idealisation, (i.e. the world full of meaning and significance as it is given to us in immediate experience), ... we measure the life—world, the world constantly given to us in our concrete world—life for a well—fitting *garb of ideas*, that of the so—called objectively scientific truths" (Ibid, pp.50—1).

Nor do we have to search far for the motive for such activity. Indeed as Heidegger's student, Marcuse, elaborates, "Galilean science ... experiences, comprehends and shapes the world in terms of calculable, predictable relationships among exactly identifiable units. In this project, universal quantifiability is a pre—requisite of the *domination* of nature. Individual, non—quantifiable qualities stand in the way of an organisation of men and things in accordance to the measurable power to be extracted from them" (Marcuse, p.134). Thus while a self—conscious pragmatism as such was of course a later development, Objectivism and pragmatism were related from the beginning. Nor should this surprise us, for, in face of a world that is, as we have seen, misconceived by impure reflection to be a threat rather than a support, it is understandable that the apparently beleaguered subject should seek to control and dominate nature by replacing all trace of qualitative meaning and significance by the theoretical constructions of an objective science.

Now, so long as all this is recognised, so long as we see the world of modern science as the subjective abstraction it is, and do not confuse it with the real world given to us in our immediate experience, then such a conception is wholly acceptable, and indeed can be pragmatically useful in enabling us to predict the behaviour of, and therefore harmoniously co—exist within, the world. The danger, however, is that we may come to confuse this pragmatically conceived, dry, lifeless, bloodless, theoretically abstracted product of Enlightenment *culture* for the *natural* world; and this indeed is precisely what has happened.

VI The Eclipse of Meaning and Value

Thus following Bacon's claim that "... knowledge ... and ... power meet in one", that "truth and utility are here one and the same thing", we find from the late sixteenth century onwards that the theoretical abstractions of the pragmatic sciences are increasingly mistaken for the real world. As Habermas notes, "The reified models of the sciences migrate into the socio—cultural life—world and gain objective power over the latter's self—understanding" (Habermas, 1971, p.113) with the result that, in Dilthey's words, "... the lively feelings in which we enjoy nature increasingly recede behind the abstract apprehension of it according to relations of space, time, mass and motion" (Dilthey, Vol.7, p.82).

Moreover, this value—free world of the so—called objective sciences is so unlike the world as it is given to us in immediate experience, full of meaning and significance, that its form stands as eloquent testimony to the abstractive activities and values of that very subject referred to by Marcuse as "... the hidden subject of Galilean science" (Marcuse, 1968, p.134) which it is at such pains to exclude from its content. As Max Scheler explains "To conceive of the world as value—free is a task which men set themselves on account of a value: the vital value of mastery and power over things" (Scheler, quoted by Leiss, 1974, p.109) Be that as it may, nevertheless as Habermas points out, "... the objective self—understanding of the sciences ... suppresses the contribution of the subjective activity to ... knowledge" (Habermas, 1978, p.212) The result, as Koyre has observed, is that modern science substitutes "... for our world of qualities and perceptions, a world in which we live, love and die, another world, the world of quantity ... in which, while there is a place for everything, there is no place for man" (Koyre quoted in Cohen, 1980, p.5).

In other words, we come here to the crux of the matter. It is precisely these reified subject—suppressing models of positivistic science, the product of the subject's strategy for overcoming the *apparent* threat from a world that in reality founds and sustains it, that, paradoxically enough, constitute the real threat to the subject. Echoing our myth, it is modern science which, in hiding or supressing all trace of the subject in its attempt to overcome our *apparent* alienation from the world, constitutes the *real* threat and the real source of our increasing alienation.

Indeed, it may be said that "Objectivity" is a mode of alienation, and lest we be tempted to turn this suggestion aside as preposterous, we should ask how else we might characterise an epistemology that replaces the life—world as given to us in immediate experience as full of meaning and significance, a world in which we actively participate, with a "picture", or passive spectacle from which the "spectator", along with all hint of any meaning and significance, has been rigorously excluded. Surely this "Objective Wasteland", is none other than the world which Camus described so brilliantly in *The Outsider*

Such a world, from which all trace of the subject, together with all intrinsic meaning and significance has been ruthlessly supressed, is no longer seen as a "home" or foundation, but is perceived instead, by the now apparently ungrounded subject, as a threat which therefore not only may, but indeed seemingly must, be exploited and subjugated if we are to survive.

VII   The Metaphysics of Objectivity

Alone and vulnerable, humans have historically sought solace and salvation through God. Modern humanity however increasingly turns to science and technology for help, while the scientific "priesthood" now receive much of the reverence formerly afforded to the clergy. And just as God can only be a source of solace to one who rejects the possibility that it is "Man who makes God", so scientific objectivism can only be effective if it too suppresses all trace of the subjective activity which, as we have seen, lies at its very heart. It is, however, as we suggested earlier, precisely in hiding itself that the subject opens itself up to danger, for, just as a belief in God *may* give rise to a spiritual complacency, to an eclipsing of our own responsibilities for our spiritual salvation which may result in our damnation, so the belief in a Scientific Objectivism, in suppressing as it does all trace of the subject, and even the subjective activity of abstraction which produces it, may pave the way for modern technology, which arguably poses the greates threat to our survival.

VIII   Scientised vs Craft Technologies

"The modern physical theory of nature", Heidegger tells us (1977, p.22), "prepares the way ... for the essence of modern technology". This, in contrast to traditional or craft technologies, which are based on practical "know−how", is a "scientised" technology, based on the abstractions of modern "objectivistic" science.

Not only does this mean that modern technology is more powerful than craft technology, but it also means that it is less restrained, for being predicated on a view of the world from which all trace of the subject has been removed, not only does it seem free to exploit such a world with impunity, but insofar as the world is actually already viewed as a threat to our now apparently ungrounded subjectivity, we feel obliged to subjugate it, lest it subjugate or overcome us. In such circumstances, it is only to be expected that, as Heidegger has observed, "The revealing that holds sway through modern technology does not unfold with a bringing forth in the sense of *poesis*. The revealing that rules in modern technology is a challenging (*Herausfordern*)". (Ibid, p.14). It is however precisely our utilization of such technology in an attempt to exploit or subjugate the apparently threatening world, which, paradoxically, poses the greatest threat to our survival.

IX   The Hermeneutics of Alienation and the Master/Slave Dialectic

Nature, no longer conceived of as a web of closely interrelated functions of which we ourselves are a part, a web of life that must be preserved at all costs if we are to survive, is perceived as a hostile threat that must be mastered. Failing to recognise that we ourselves are part of nature, we therefore, like some latter day Dorian Gray, fail to perceive the toll that our exploitation of nature is exacting from us. Indeed we even come to regard the ever more efficient extraction and processing of non—renewable resources and the endless stream of polluting, toxic and life—threatening liquids and gases belching from our industrial plants as a symbol of success. Further "wildness" is "tamed" and turned into housing tracts that can never be appropriately called "home", while cities, often perceived as monuments of our self—affirming mastery of nature, are, for many, "concrete jungles" that threaten their citizens' health and sanity, and indeed their very lives. The possibility of community (characterised by Scheler as a unified integrated totality guided by "brotherhood"), remains unrealised, and we have instead a plurality ruled by laws, while individuals, each potentially valuable for their idiosyncratic uniqueness, increasingly become mass producers and consumers, whose activities and tastes must be "educated" (regulated) in the service of the industrial machine, be it capitalist or socialist. "In self—assertive production", Heidegger tells us, "the humanness of man and the thingness of things dissolves into the calculated market value of a market which ... trades in the nature of Being" (Heidegger, 1970, p.115). We come to know the price of everything and the value of nothing, and should our portrait, like that of Dorian Gray, be temporarily hidden by some ecological or political time—lag, we have only to look out into the world, a world of starvation in the midst of surplus, a world where cut—throat competition between individuals and between nations so often eclipses the potential for harmonious co—operation, to see the reflection of our own alienation. Truly it appears, as Hegel saw, that the masters are themselves held captive.

Nor is any of this to envisage a Golden Age, but merely to point to the fact that while even at current population levels we are capable as *never before* of eradicating starvation and world poverty, together with the cut throat competition which would be understandable, even pardonable, among those who are fighting for their very survival, in fact more people are starving than ever before, and it is precisely those nations that are comparatively well off whose arsenals of weaponry threaten, for the first time in history, all life on the planet. It is then *this every widening gap between our potential and actual achievements* that is surely the real indicator of our alienation.

X Alienation and Scarcity

Now in view of our undoubted technological *potential*, continued economic scarcity must be seen as a consequence rather than a condition of our alienation. It would therefore be wrong to conclude as many, including some Marxists, do, that the abolition of scarcity would bring about the end of alienation. On the contrary, it would seem that the obverse is true and that, as we can perhaps now see from this analysis, far from scarcity as such being responsible for our alienation it is our alienation which perpetuates scarcity in the midst of plenty. Indeed, as we may perhaps now be able to see, it is the very technocratic reduction of human being to the status of economic instrumentality, and the world to so much "grist for the mill" or "standing reserve" as Heidegger calls it, to be set upon and exploited with apparent impunity, that constitutes our actual alienation; an alienation which we have argued is rooted in the "Objective" world view of modern science and the reificatory reflection upon which it is based.

XI   Being, Doing and Having

After all, who else but economic or technocratic wo/man, oblivious of their socio—cultural rootedness in the traditional life—world, and consequently attempting to replace their lost sense of Being with economic and technocratic Doing, would feel so much anxiety at the prospect of even superannuated retirement? Who else, in the face of facts, would need to convince themselves of their absolute indispensability to their employer? Who else would feel such antipathy at the prospect of a post—scarcity economy in which much drudgery in the service of production would be replaced by self—rewarding activity? And if Doing, or producing, cannot fill the void left by the lost sense of Being, there is always that other aspect of economic and technocratic existence, having, or consuming, to convince us of our own substantiality. As Marcuse writes "... people recognise themselves in their commodities; they find their souls in their automobiles, Hi—Fi sets, split level home, kitchen equipment" (Marcuse, p.24).

We have only to look at "advanced" industrial societies, societies in which many engage in an endless merry—go—round of seemingly purposeless and trivial activity; a compulsive round of getting and spending, of neurotic production and consumption, of distraction from distraction by distraction, to see that it is not scarcity that is responsible for our alienation, and which poses the biggest threat to ourselves and our fellows and our environment, but the loss or suppressing of our own primordial sense of Being and belonging: a failure to comprehend what Heidegger referred to as our "Being—in—the—world with others".

XII   The Foundation of Salvation is the Salvation of Foundations

"The concrete life—world", Husserl tells us, "is the grounding soil of the 'scientifically true' world", and it may be that only by retracing our steps back towards this concrete life—world, towards this primordial awareness of our "Being—in—the—world with others", which, like Husserl, Heidegger also regarded as the original foundation from which all epistemological quests, including that of modern science, proceed, that we can hope to overcome our alienation and the threat of annihilation to which it gives rise.

However it is one thing to prescribe such a programme and another to describe how this "return to our origins" is to be concretely implemented. Our best hope of finding an answer is, I believe, to employ an ontological archaeology, to attempt to travel back beyond our earlier ontogenetic analysis of the reflective origins of the sense of the individual as an isolated ego, to uncover the phylogenetic foundation. For it is here that we may expect to find the ultimate source of our alienation; and it is this that we must understand if we are to reintegrate ourselves.

XIII   Phylogenesis and Alienation

The evolutionary development of wo/man is remarkable for the fact that his/her head leaves the ground; that is, that s/he walks upright. Along with this development, sight and sound, given a new vantage point, replace smell as the nose is distanced from the ground, while the body with its sense of touch is neglected and even repressed in favour of the head. The *participatory* senses of touch and smell recede into the background along with the *body*, while the *spectator* sense of sight comes to the fore along with the *head*. Thus isolated in our heads from our bodily involvement with the world, we, who now identify ourselves increasingly as a *rational* ego and decreasingly as an embodied *animal*, repress the physical in favour of the mental, the lived experience in favour of the reflective, abstract and ideal, finally forgetting that, while our heads may be in the clouds, with the gods, our feet remain firmly on the ground. As Roszak notes "... this hypercapitated picture of the body (is) the chief obstacle to organic wholeness". "... most people have taken up exclusive residence in the head ... the head becomes *headquarters*" (Roszak, 1973, p.94).

Is it possible, as Erwin Strauss appears to suggest, (Strauss, 1952, pp.529–61), that we have here the ontological ground of the epistemological reflection which we have already suggested is responsible for our feeling of alienation? Could it be that at this phylogenetic level our alienation is related to changes of the most concrete kind in our physical relations to the world and to resultant changes in the ratio of the development of the various senses hinted at above? Could it be that such *perceptual* shifts result in profound *conceptual* shifts? Let us now investigate this possibility.

XIV   Technological Mediation of Existential Relations

McLuhan notes, "The effects of technology ... alter sense ratios or patterns or perceptions..." (McLuhan, 1967, p.27) and it is our hypothesis that in doing so it will influence our picture of the world and therefore our relation to it. Indeed, McLuhan produces much evidence to this effect. For example, he shows how, in contrast to primitive societies which were permeated by "cool" media (those high in participation or involvement) (*Ibid.*, p.31) literate society, relying as it does on the visual sense, which as we have already noted is a spectator sense and hence "hot" (or low in participation) (*Ibid.*) has produced a heightened level of alienation. He notes — "Literacy ... by extending the visual power ... conferred the power of detachment and non–involvement" (Ibid., p.357).

With remarkable insight McLuhan, reiterating and further explicating what we have hinted at, concludes that — "Visual detachment ... gave the power of the second look — the moment of recognition. This broke people out of the bondage of the uncritical and emotionally involved life. It also fostered the cult of private competition ... for private power ... " (McLuhan and Nevitt, 1972, p.39).

Nor is this power of detachment exclusively conferred by the print technology of the "Gutenberg Galaxy"; television also, while extending our consciousness of the "Global Village" as never before, substitutes a cerebral image for the total sensorium of bodily involvement. For example, while we may all have seen TV pictures of the moon we do not actually *feel* we have been to the moon. The moon, the latest war, starving millions, mass executions, polluted ecosystems, energy crises, etc., all become part of an endless spectacle from which we are somehow distanced as non—participants. While it would be wrong to conclude that such images have no effect upon us — charities raise large amounts as the result of TV pictures of starvation in the Third World, for example — nevertheless such cerebral images stand in contrast to participatory involvement, and one cannot but wonder how much more would be raised if we had all been and experienced at first hand the disasters reported to us.

For exactly parallel reasons the electronic battlefield with its computer controlled visual display has opened up the all too real likelihood that someone who would find it emotionally impossible to kill even an animal with a knife, let alone hack off the limb of an adversary with a sword, will find it all too easy to unlease a thermo—nuclear device which will kill hundreds of thousands, or even millions, of men, women and children, many in a slow and painful way, by burning or frying them to death, or producing long term cancers, at the press of a button. As Ellul confirms "... Technical progress favours wars ... because the new weapons ... have enormously reduced the pain and anguish implied in the act of killing" (Ellul, 1964, p.110).

Nor is the technological mediation of the relationship between us and our world limited to sensory devices. It is, after all, commonplace to observe how other non—sensory forms of communications technology, such as aeroplanes and cars, have "shrunk the world". We should perhaps note that this shrinkage is far from uniform. For instance, to fly from New York to London takes me only one and half times the time it takes the bus to drive me the 200 or so miles from the airport to my home, while in the hour and a half it takes me to get the first 30 miles on the congested roads near the airport I could have travelled almost 1000 miles by airliner. If we were, therefore, to construct a phenomenological map *spatially* scaled according to my *temporal* experience of travel we would find ourselves in an almost unrecognisable world. We may experience analogical distortions of the life—world due to selectivity of the media; we may, for example, find we know more about interplanetary flight than about how to grow the food we eat, or about "how the West was won" than about current world affairs; while the selection of news items, for example, defines for us what we will consider to be significant events or topics of concern.

Important and far ranging as all these effects of technology undoubtedly are, there are yet even more fundamental ways in which technology can influence our sense of participation in, or alienation from, the world. Let us take for instance the experience of driving down a country lane at fifty miles per hour in an air—conditioned car, perhaps listening to the radio. This is quite a different experience from a leisurely stroll down such a lane. While the driver may be focussing on the road some yards ahead and may only be dimly aware of the trees and the fields constituted in his/her peripheral vision as horizons of his/her experience, the walker, who a few minutes before may have been in exactly the same physical space as the driver, may have smelled leaf mould, stepped in a cow pat and been drenched by the rain, and upon rounding a bend s/he may have been confronted by an uprooted tree, the victim of a recent storm, lying across the road. Enthralled by the lichens and mosses growing on its side and the ivy entwined round it s/he may have stopped to take in the spectacle in full, suddenly to become aware of a car rounding the bend and an infuriated driver slamming on his/her brakes and hooting at the impediment to his/her passage. Truly then in the words of the poet Blake "A tree which can move one man to tears, for some man is a green thing that stands in the way".

It would be wrong to conclude that walking is therefore in some sense better or more worthy than driving. After all, our driver may be aesthetically uplifted by the concert being broadcast on the radio, and is almost certainly more comfortable than our walker. The point is simply that, as Sartre has argued, the instruments we employ define what Sartre calls the "instrumental complexes" and "hodological spaces" that permeate our life—world" (See, eg., Sartre, 1956, pp.321—2).

Like the car driver, city dwellers are also physically insulated from the countryside. Living in air conditioned apartments and automobiles they are insulated from the weather, from the primordial organic rhythm of the seasons, and even, by the power of the electric light, from the rhythms of night and day. It is perhaps not surprising under such conditions that cyclical time should be eclipsed by a Newtonian linear temporality, while seasons are marked only by a change in fashion as the clothes designers bring out their Autumn collections; and that, no longer perceiving ourselves as part of the endless organic cycle of life, we should conceive ourselves as a point whose distance from its end is, in Eliot's memorable phrase, "measured out in coffee spoons".

Technology therefore has a definite intentionality, for in so far as we tend to "focus" most strongly on those aspects of the totality which present the greatest possibilities for or potential threat to our projects, and in so far as technology in mediating our relationship to the environment thereby defines these possibilities and threats, then technology affects what we will "focus" on and consider significant. Thus even had our driver not been isolated by the radio and air conditioning from the sounds and smells of the country s/he would probably none the less have remained to a large extent oblivious of them, repressing the other senses in favour of the sense of sight, and even adjusting the visual field to the task at hand. Moreover, in addition to influencing our *perception* of the world, technology also influences our *conception* of it, as is exemplified by the different conceptions that our driver and walker had of the same tree. At a more profound level, the Cavalry Officer, for example, born of a predominantly agricultural technology, can neither see nor structure the signs left by his quarry, and calls upon the Indian Scout or Backwoodsman, born of hunting technology, to open up the life—world of the hunter for him. We can see then how the advent of modern technology has not only opened whole new worlds of underwater and space exploration, particle physics and astronomy, but, in demanding of us that we focus in just such and such a manner on such and such "facts", while allowing other previously indispensable facets of existence to drift off into areas of peripheral concern, and in demanding that we structure our experiences in such and such a way, modern technology has opened up the old world in totally new ways.

Now not only are our experiences technologically mediated; the myths and models in terms of which we go on to interpret these experiences are themselves derived, at least in part, from the technosphere. Thus animistic world views deriving from hunting and agricultural technologies gave way to the Newtonian clockwork universe, with its metaphor of the body as a machine, derived from the mechanisms of the seventeenth century, while these in turn gave way to the electronic and cybernetic images of the twentieth century, in which the brain is regarded first as a telephone exchange, and later as a computer. Indeed even our modern heroes, from Superman onwards, reflect in mythical image form the potency of our technology; while science and technology, trying to live up to their own myths, send humans into space, thereby producing heroes of their own whom we are encouraged to emulate. Could it be that we who stand alienated from our ownmost founding origins seek to substitute an "outer" voyage of conquest for an "inner" voyage of discovery? Could it be that this physical conquest of the heavens is, in part at least, a substitute for a more primordial quest?

Meanwhile the education system, ever more closely linked to the labour requirements of the industrial society, be it socialist or capitalist, encourages us to perceive and project ourselves in terms of the increasingly fragmented work roles of a more and more specialist division of labour, while in the long term developments in technology, by influencing and molding our environment, may even determine our evolution itself (See Hall, 1969). Closure is fast approaching.

XV One Dimensional Man

We can see then that technology mediates at every level our perceptions, experience, conceptions and interpretations of ourselves, others, and the world, and that in so doing it directly determines our sense of, and our relations to, the world. The choice of technology is, therefore, crucial to any attempt to overcome our alienation. However, and here is the apparent dilemma, it is against the background of a world — image and a self — image that are mediated at all levels by the prevailing technology that we, who are ourselves ultimately the product of an evolutionary process that is itself even now to some extent at least informed by the technosphere, must choose our technology. Therefore, while our technology is ultimately our choice, we have grown up within a technosphere which in the manner just outlined has already mediated not only the ratio of development of our senses and central nervous system, and consequently determined *how* we experience, but also mediated the very way we focus our attention, and therefore even *what* we experience; a technosphere which, as we have seen, even presents us with the symbols of success that will be so important an influence on the values that will guide our choices between those very alternatives and possibilities that it delineates. Small wonder then that, as Marcuse claims, increasingly in "... advanced industrial society ... the technical apparatus of production and distribution ... functions, not as the sum total of mere instruments, but rather as a system which determines *a priori* the product of the apparatus as well as the operations of servicing and extending it. In this society, the production apparatus tends to become totalitarian, to the extent to which it determines not only socially needed occupations, skills and attitudes, but also individual needs and aspirations" (Marcuse, 1968, p.13).

In this manner, modern scientific technology is instrumental in defining the telos that in turn forms the basis of our choice of technology. Alienation instantiates itself in a technology which hermeneutically reproduces the condition of its own creation. Means increasingly become ends as, following Hegel's master/slave dialectic, we become the servants of our own creation, the product of our products.

While many of us find such conclusions unpalatable, the facts are, I am afraid, unambiguous on this point. Think for example of the arms race, a race with no winners, whose most successful possible outcome can only be the waste of vast amounts of money, skill and effort that could be far more profitably spent upon eradicating world hunger. Both sides acknowledge this, yet both sides appear to have been trapped, the slaves of their very technological "success", into an open—ended commitment of the most alien kind. Indeed it is perhaps not overstating the case to say that most of the global problems that confront us, the threat of thermo—nuclear annihilation, pollution, resource depletion, etc., etc., are not the result of failures in our science and technology, but the products of their very success. Following this process, every "technological fix" will, so long as it is predicated upon "objectivistic science" simply instantiate or extend the process of alienation, of which it is symptomatic, on another level.

XVI Conclusion

With quite remarkable insight, Jaspers sums up. "Man's attachment to nature is made manifest in a new fashion by modern technology. Nature threatens to overpower man himself, in a manner previously unforeseen, through his tremendously increased mastery of her ... The danger threatens that man will stifle in the second nature, to which he gives birth technologically as his own product, whereas he may appear relatively free *vis à vis* unsubdued nature in his perpetual struggle for existence. Technology has wrought a radical transformation in the day by day existence of man in his environment; it has forced ... the metamorphosis of his whole existence into a technically perfect piece of machinery and of the planet into a single great factory. In the process man has been deprived of all roots. He is becoming a dweller on the earth with no home" (Jaspers, 1953, p.98).

Now if, as we have argued, our conceptions, and indeed even our very perceptions, are increasingly subject to a technological mediaton which is fast approaching "one—dimensionality or closure", then it would seem that escape is impossible. However, Heidegger, quoting Hölderlin, tells us:

> " 'But where the danger grows,
> The saving power also'" (Heidegger, 1977, p.28).

From where then does "the danger" which "consists in the threat that assaults man's nature in his relation to Being itself ..." (Heidegger, 1971, p.117) grow? From this very relation, of course, as we have seen; and it is therefore clearly to this relation that Heidegger is directing our search after salvation.

Now I have argued at considerable length throughout that our alienation is the outcome of the objectivistic suppression of the subject in response to an apparent threat supposedly emanating from the world, a world which is, in reality the ground or foundation of our very existence. This ontogenetically evolved alienation itself in turn grows out of our prehistoric phlogenetic evolution, from the rational detachment from, and suppression of, our lived experience of and relation to the world. It is therefore ultimately this ontologically primordial lived experience of, and this original relationship to, the world, this absolute point of origin of our alienation, to which Heidegger thus suggests that we look for salvation. Indeed, if our "one — dimensional" closure thesis is correct then we are denied any *external*, "Archimedean point" or perspective from which to mount a critique of technology. This being the case, it is clear that it is only in so far as we are able to draw upon such primordial lived experience that we are able to effect the distancing which, as Habermas has rightly insisted, is the indispensable precondition of critical theoretical reflection.

However, the belief that theory and theory alone can save us itself symbolises precisely that rational detachment from and suppression of our original lived experience of the world, so symptomatic of our alienation. It is therefore clear that we can only be saved by a praxis that combines Theoria, the "dis — covering" of our grounding origins, with a technological practice that, at the pre — reflective level of lived experience enhances, rather than suppresses, this relation to, and dependence upon, the world and others.

As we can now see, only a species that wants to be self — founding Lords of Creation, and thus free from dependence upon any other foundation, could come to regard their actual grounding and sustaining foundation as alien, while clearly the attempt to conquer such foundations, far from providing liberation, is an act of self — enslavement that may yet lend to our destruction. Thus as the Eden myth demonstrates, our best hope of salvation lies NOT in attempting to transcend the limits and prohibitions emanating from our grounding and sustaining origins, but precisely in accepting them as the precondition or foundation of our existence.

As Cerezuelle has pointed out, "When we no longer expect technology to provide false liberations and an impossible transcendence of the limitations that are inseparable from the organic condition of our existence, we will no longer allow it to develop in a totally devouring way" (Cerezuelle, quoted in Durbin, 1979, p.72).

It is thus upon a praxis that combines a thinking of grounds with a grounding of thought that our salvation depends.

References

Cohen, *The Lineaments of Mind*, W.H. Freeman, Oxford, 1980.

Dilthey, W., *Gesammelte Schriften*, Vandenhoeck and Ruprecht, Gøttingen, 1913—67.

Durbin, P. (ed.), *Research in Philosophy and Technology*, Jai Press, Greenwich, Conn., Vol.II, 1979.

Ellul, J., *The Technological Society*, trans. J. Wilkinson, intro. R. Murton, Random House, New York, 1964.

Freud, S., (1) *Civilization and its Discontents*, W. Norton, New York, 1962.

Freud, S., (2) *The Ego and the Id*, trans. J. Riviera and J. Strachey, The Hogarth Press, London, 1963.

Habermas, J., *Toward a Rational Society,* Heinemann, London, 1971.

Habermas, J., *Knowledge and Human Interest*, 2nd ed., Heinemann, London, 1978.

Hall, E.T., *The Hidden Dimension*, The Bodley Head, London, 1969.

Heidegger, M., *Being and Time*, trans. J. Macquarrie and E. Robinson, Harper & Row, New York, 1962.

Heidegger, M., *What is a Thing?* trans. W. Barton & V. Deutsch, H. Regnery, Chicago, 1967.

Heidegger, M., *Poetry, Language, Thought*, trans. & intro. A. Hafstadter, Harper & Row, New York, 1971.

Heidegger, M., *The Question Concerning Technology and Other Essays*, trans. W. Lovitt, Harper & Row, New York, 1977.

Husserl, E., *Phenomenology and the Crisis of Philosophy*, trans. Q. Lauer, Harper & Row, New York, 1965.

Husserl, E., *Cartesian Meditations*, trans. D. Cairnes, Martins Nijhoff, The Hague, 1970.

Husserl, E., *The Crisis of European Science and Transcendental Phenomenology*, trans. & intro. D. Carr, Northwestern University Press, Evanston, Ill., 1970.

Jaspers, K., *The Origin and Goal of History*, trans. M. Bullock, Routledge & Kegan Paul, London, 1953.

Krell, D. (ed. & trans.) *Martin Heidegger: Basic Writings*, Harper & Row, New York, 1978.

Leiss, W., *The Domination of Nature*, Beacon Press, Boston, 1974.

Marcuse, H., *One Dimensional Man*, Sphere Books, London, 1968.

McLuhan, M., *Understanding Media*, Sphere Books, London, 1967.

McLuhan, M. & Nevitt, B., *Take Today*, Harcourt Brace, New York, 1972.

Roszak, T., *Where the Wasteland Ends*, Faber & Faber, London, 1973.

Sartre, J—P., *Nausea*, trans. L. Alexander, New Directions, London, 1949.

Sartre, J—P., *Being and Nothingness*, trans. & intro. H. Barnes, Philosophical Library, New York, 1956.

Sartre, J—P., *The Transcendence of the Ego*, trans. & intro. F. Williams & R. Kirkpatrick, Farrar, Straus & Giroux, New York, Undated.

Strauss, E., "The Upright Posture" in *Psychiatric Quarterly* XXVI, 1952.

4 Technology and medicine: means and ends

HARRY LESSER

That advances in technology have made medicine vastly more efficient is obvious. What is less obvious is the extent to which it has changed the actual aims and purposes of medicine, the ways in which, by changing what is possible, it has also changed what people can and do aim at, as well as changing what they want, need, or think ought to be done. In this paper, I want to consider the nature of some of the more important of these changes.

It might be objected that fundamentally the aims and essence of medicine have remained constant over the centuries. Has not the aim always been to promote health and prevent or cure disease? Moreover, within the medical art or science, one can roughly distinguish the task of keeping someone safe and comfortable while nature takes its course — i.e. nursing — from that of tackling specific problems with specific remedies or attempted remedies, which is medicine (The use of these terms is not meant to imply anything about the proper roles of nurses and doctors). On this definition, the essence of medicine lies in 1) diagnosing the nature of the patient's problem, which involves identifying either the malfunctioning which is causing the symptoms or the particular classifiable syndrome of which they are part, followed by 2) trying out one or more remedies designed to correct the malfunctioning or end the syndrome. And so, someone might suggest, the *essential* nature of medicine has not changed, nor have its aims.

On a very general level this is true. But what does change is the set of problems that are seen as being medical. Improved technology has made it possible to use medical techniques where previously they had no function, which has meant that what was in earlier times either not a problem at all (what one can do nothing about is not perceived as a problem), or at any rate not a medical problem, has now entered the medical sphere. This in turn alters our notions both of what constitutes being healthy or   unhealthy

— since a medical problem is a health problem — and also of what is or is not a doctor's business. The verbal statement of the aims of medicine does not change, but the 'cashing out' of those aims does.

One may give three examples. The first is having children: as long as there was no scientific way of making the infertile fertile, it was no part of medicine to try to do it. Now, however, the cure of infertility is an accepted part of medicine (though there are ethical problems as to what methods of cure are justifiable); and infertility itself is, though not an illness or disease, a health problem of a kind. And if we go back in history, one can well argue that originally midwifery was a craft separate from medicine, and that it entered medicine only with the appearance of specifically medical techniques for making pregnancy and childbirth easier and safer.

A second example is the process of aging. Improved medicine has resulted in some conditions, such as arthritis or weakening of teeth, as being generally seen as unhealthy and requiring treatment, whereas in the past they were often regarded, at least by the poor, as being merely the inevitable accompaniments of growing older. Moreover, the extension of the aims of medicine to include enabling people to retain physical vigour as long as possible is generally accepted; and although little progress has been made in extending the human life span (as opposed to helping far more people to live it out), this also is often accepted as an appropriate area for medical research.

Thirdly, there is the relief of pain. It would seem that, while no doubt doctors have always tried to avoid unnecessary pain, in the past, when little could be done medically to relieve pain, they saw their real function as healing injurious conditions rather than ending discomfort. There is some evidence that this attitude can still prevail: some doctors still see the use of some pain killers as to be avoided as an 'interference with nature', a view they would hardly take of, for example, antibiotics. But in general, it is seen as a function of medicine to try to remove pain and discomfort as such, not merely as an incidental consequence of healing.

The consequences of changes such as these are various, but interconnected. First, there is the change in the standard of 'health'. Essentially, the standard has been raised: more conditions are seen as detracting from health and as something that ought to be cured, fewer as mere consequences of being old or poor or unlucky or simply human. Also, some conditions are perceived as being unhealthy at an earlier stage or a lower level: pains do not have to be so intense to be legitimate matters for concern, and defects, for example of sight or hearing, do not have to be nearly as pronounced to count as defects.

This in its turn affects people's expectations, both of what they think will happen and of what they think they are entitled to. Largely, people from the better—off sections of advanced industrial societies expect to stay well, at the new higher level of being well, for most of the time, and to be cured, or at least have their discomfort or disability eased, when they are ill. This may also be seen, consciously or unconsciously, as a right, not just to be treated, but to be cured. This sense of having a right to be cured may be reasonable or unreasonable — some writing in the Press on the 'scandal' of the number of deaths from, e.g., heart disease seems to ignore the fact that we must all eventually die of something or other. But the crucial point is that, because what is possible has changed, our notion of what we have a right to has changed — not always in step with the changes, but as a result of them.

The change in claims and expectations, together with the improved technology itself, has had various effects on standard medical practice. Thus, there seems to be a tendency for the availability of equipment to encourage its use. One obvious area is pregnancy and childbirth, where the use of technology steadily increased until a sizable number of people began, rightly or wrongly, to question whether it was beneficial. In general, one might suggest that there has been a change in style, a change that has produced less nursing, i.e. providing comfort and letting nature take its course, and more medicine, i.e. diagnosing and intervening in order to put right or improve a condition.

One might sum this up as follows. Both practitioners and patients can take views of medicine that vary in scope. At one extreme, there is the 'narrow' view, that medicine, i.e. diagnosis plus intervention, is appropriate only to a limited range of human problems, involving specific diseases, injuries and unhealthy conditions: other problems, even if physical, require nursing, counselling, or simply endurance and patience. Moreover, even within this sphere there can be no strong expectation of success, much less any entitlement on the part of the patient to be cured — his only claim is that the doctors do their best. Nor is there any feeling that the area of medicine should be extended beyond what it currently handles.

The contrasting 'broad' view sees the aim of medicine as that of diagnosing and putting right as much as possible: the only reason for holding back is lack of technical expertise, and there is always the hope that this will be rectified by more research. Failure, however common, and whether in the application of existing remedies or in discovering new ones, is in a sense a disgrace: it may even be seen as a failure to meet reasonable expectations and entitlements. There are still limits to medicine, but they are to be pushed back rather than accepted.

These views are on a continuum: one can have views of varying degrees of 'breadth'. In all probability this continuum has always existed: whatever the stage of development of medical science at any particular time there have probably been 'maximalists' and 'minimalists' and those in between. But the effect of the great technological advance in medicine has obviously been both to extend the scope of medicine and to encourage its maximal use; and only fairly recently has the desirability of this been at all widely questioned.

For the merits of the maximal view of medicine are obvious: lives saved, injuries healed, pain relieved, epidemics halted. There are, however, as various people have pointed out, disadvantages as well. First, there is the problem of unreasonable expectations — admittedly, a problem caused by patients themselves rather than by the medical profession; but nevertheless a problem. As members of the animal kingdom, we are all doomed to die sooner or later; we are all doomed to lose some of our faculties with age (unless we die first); we are all doomed to experience physical suffering (though some of us are much luckier than others). Medicine can reduce the risk of an untimely death, and possibly prolong the life span a little; it can slow down and ease the disabilities caused by age, and it can relieve pain and suffering; but it cannot ever remove death, aging or discomfort. Hence the unreasonable expectation that there can always be a cure, or the subtler and worse mistake of imagining that one has a right to be well or to be cured, add to the physical suffering the mental pain of needless disappointment and resentment.

Secondly, there is the emphasis put on trying to cure or to use technology even where the likelihood of success is infinitesimal or zero, at the expense of the patient's comfort or dignity. One example is the use of technology in childbirth: some uses of it obviously increase the child's chance of survival, others ease the mother's pain and discomfort; but sometimes the mother's fear, discomfort and sense of isolation are all increased by the atmosphere of 'high technology', with no countervailing benefit for the child. Another example is the distress caused to the dying by measures that prolong life for a little at the cost of making it not worth living: one striking feature of a recent programme on AIDS was the suffering and helplessness of *some* of those treated by orthodox medicine, compared with the cheerful resilience of *some* treated by 'alternative' methods — who may live no longer, but much more happily.

A third example would be those situations where medically the best thing to do would be to let matters take their course — it is not, obviously, for the layman to say when this happens — but where patients' expectations that something will be done force the doctor to take some action, even though he regards action as unnecessary, and conceivably a little harmful. In all these cases comfort, well—being and dignity are sacrificed to an attempt at intervention that produces no, or very little, benefit; and the reason for this is the ideological conviction that all that can be done should be done, even if it does no good. It should be added that this belief, and the harm done by it, can be the responsibility of patients as much as, or more than, practitioners. Also, to replace it by a blanket anti—technology view would clearly be to throw out the baby with the bath—water: what is needed is a readiness to take each case on its merits, without pre—conceived ideology.

The third problem is the extension of the medical model — i.e. diagnosis plus intervention — to cases for which it is inappropriate. Some of the examples above are instances of this, as well as of disregard for comfort. There are also occasions when patients look for a cure when what is really needed is a change in diet or life—style or their general situation. But the biggest problems here perhaps occur in psychological, rather than physical medicine. For on the one hand the medical model is widely used in this area, and by psychotherapists and clinical psychologists as well as by psychiatrists: it is not simply the result of a medical training but of a general belief that problems can be cured by identifying what is wrong with the person who has the problems and intervening, physically or in some other way, to try to put it right. On the other hand, many psychologists (to use a general term) have argued that these problems are in no sense medical, and ought not to be dealt with by therapy at all.

The most radical expression of this view is in the series of books by Thomas Szasz (for example, Szasz, 1974). It can also be found in the work of R.D. Laing (eg. Laing, 1965), and more recently in the work of E.K. Ledermann (Ledermann, 1984) and David Smail (see Smail, 1984, and the article in this volume). If one ventures to put together insights from all these writers, all of whom are also practising psychologists or psychiatrists, one can reach the following position, which probably none of them would actually hold, but which serves to show the nature of the problem. (It is, indeed, based essentially on the view of Szasz, who is the most radical of the four.)

The position is this. Only what can be traced to physical malfunctioning, disease or injury can be properly called an illness: hence, as Szasz would put it, 'mental illness' is either a special kind of physical illness — a malfunctioning of the brain or nervous system — or it is not an illness at all. Moreover, the scope of the term 'illness' may be further reduced by a scepticism about attempts to demonstrate the physical causes of mental 'disturbances', and a belief that this is in many cases so far unproved: the evidence that schizophrenia is physically caused or influenced makes it, for Szasz, at this stage still only a "putative disease" (see Jonathan Miller's interview with Szasz in Miller, 1983, p.282).

These conditions which cannot be classified as illnesses require to be reclassified in one of a variety of ways. They may be mere 'eccentricities', types of behaviour which are unusual and to some disturbing, but which either do no harm or do harm only to the individual and only of a kind which he himself either does not regard as harm or has consciously chosen to find acceptable — so that there is no justification for other people to interfere. They may, however, do harm to others or violate their rights, in which case, on Szasz's view, they should be openly and honestly treated as criminal behaviour, and not as illness. Or they may be 'problems in living', either existential problems within the individual, which are explored by Szasz and Ledermann, or problems arising out of their social situation, which are explored by Laing and Smail. These are genuine, often grave, problems: but they need to be dealt with by counselling or by social action, not by the application of a pseudo—medical procedure.

In all these cases, the problem is not that of an individual functioning badly, as in illness, but of an individual in a bad situation. A bad situation cannot be cured by intervening to put the person in it right. Such intervention can, it is true, sometimes make the situation tolerable — eg. by the provision of tranquillisers — and sometimes this is the only thing that can be done, or is a necessary prelude to anything else: the person may need to be calmed down before they can begin to 'take stock' and think of ways of improving matters. But, however necessary this may sometimes be, it cannot solve the real problem; and sometimes, by making the person reconciled to their situation, rather than actively resisting it and doing something about it, it may make matters in the long run worse.

Also, while it may well be possible to offer advice or counselling which can enable someone to improve their situation, this counselling, even if carried out by someone medically qualified, is essentially different from the practice of medicine. This is true not only of the general giving of advice, but also of the use of relatively formalised and, in a sense, esoteric techniques for obtaining self—knowledge, such as psycho—analysis: the 'patient' is not being 'cured' by the technique, but helped to understand what is happening and why, in order that their control over their own destiny may be increased. Essentially, there is the work of removing undesirable physical conditions, which is medicine, and for which the question of the patient's autonomy is irrelevant, except that, as a fellow—human being, the doctor may hope to increase their autonomy by freeing them from malfunctioning. On the other hand, there is the work of helping people to understand themselves and their relations with other people, in order that they may be able to make choices that are better (in their own eyes) and more genuinely autonomous, more in line with what they themselves consider best for themselves and other people. To call this medicine, for those who hold this position, confuses the issue, treats people with insufficient respect for their ability to make choices and look after themselves, and may result in the wrong things being done, particularly in attempts to alter the individual instead of the situation.

A full assessment of this position requires medical knowledge. But even from a layman's point of view various points can be made. First, the distinction is a valid one, and therefore the possibility of a problem being wrongly treated as medical certainly does exist; and evidence from, eg. David Smail (1984), shows that it is not merely a possibility but a regular occurrence. On the other hand, though, evidence from Ledermann (1984) shows that the opposite mistake, that of treating a person as making autonomous choices when their behaviour is in fact the result of a physical condition, also occurs, though perhaps less often, and does just as much harm.

Indeed, accepting the validity of the distinction should not lead us to draw it precisely where Szasz, for instance, does. There are three reasons for this. First, Szasz is probably unduly reluctant to admit physical causes even where there is evidence for them: schizophrenia is only a 'putative illness'. But surely, it ought to be enough that there is good evidence that a physical disturbance is part of the problem, always provided that there is an adequate probability that physical treatment will do good rather than harm — to insist on proof is surely to require too much. It is also quite possible that, as medical knowledge increases, so knowledge of the physical causes of mental disorder will increase, and therefore this area will grow. However, it still needs to be emphasized that, given the conceptual arguments above, it is inconceivable that it could take over the whole area of human problems, as a few people seem to imagine: all that can be said is that there is likely to be *some* increase in what can be attributed to physical causes.

Secondly, although the distinction is a valid one, actual situations are very often going to involve both a medical problem and a problem in living; and although the two require a different approach, both approaches need to be combined in order to deal with it. Thus the person in question may be, among other things, simply physically ill, whether or not the illness is psychosomatic in origin. Or, as mentioned above, they may need to be brought down from an over—excited state, or out of a depression, by physical means before it is possible for them to tackle what they are actually depressed about. What is needed here, it would seem, is awareness that there is not *only*, and perhaps not primarily, a medical problem.

Thirdly, *pace* Szasz, there are various conditions which, though they may have no known physical cause and are not strictly speaking illnesses or diseases, *resemble* physical illnesses much more than they do autonomous choices; i.e. they are involuntary conditions which interfere with a person's general functioning, not decisions to behave in an unusual way. Physical conditions are rightly regarded as the paradigm cases of illness; but this does not mean that any extension of terms such as 'illness' and 'disease' beyond the physical sphere is *ipso facto* improper. Essentially they become appropriate when applied to conditions that need to be *treated* as illnesses or diseases, rather than crimes or eccentricities.

One obvious example would be delusions or hallucinations. There is, it is true, a problem in deciding which beliefs are delusive, which are false but rational, and which are simply a disagreement with the views of the majority, the establishment, or the therapist; there may, after all, be Soviet psychiatrists who genuinely believe that political dissent is proof of mental illness. Nevertheless, the point must come when the likelihood of a belief's being false is so high, its grounds so weak and the consequences of acting on it so terrible that it should be treated, on the grounds of humanity, as a delusion. It is true that one could still argue that having a false belief is not *really* a medical condition; but from a practical point of view the person in question neeeds to be treated *as if* they were ill, rather than as if they were making an autonomous choice. A good example would be the beliefs of a serious anorexic that they are not underweight, that they are in physically good condition and that they can continue their extremely low food intake and still remain alive: the equation by Szasz of anorexia with a political hunger—strike (Miller, 1983, pp.287−8) simply ignores the influence of these delusions and the fact that they make the choice not to eat no longer autonomous. Szasz's analysis may well have explanatory value, but should not affect practical treatment of an anorexic in danger of starving to death.

A second group of examples would include a variety of conditions which are similarly involuntary, detrimental to efficient functioning, and, so far as one knows, not always physically based. These include clinical depression, whether reactive or endogenous, manic states, extreme inability to concentrate, obsessions, compulsions, phobias and extreme anxiety. Again, it will presumably sometimes be uncertain when one has a perfectly intelligible emotional reaction to an intolerable or near—intolerable situation, when one has, effectively, an 'illness' or malfunctioning that needs to be cured, and when a person's problems involve both. But once again one can say that sometimes these conditions, even if not physically caused, are sufficiently serious for the person to require the medical mode of help, whatever counselling or support or encouragement towards autonomy they may need at the same time or thereafter.

A reasonable conclusion might be this. There are conditions, not all of them physically based, which amount to an involuntary malfunctioning, and for these a medical approach is appropriate. There are other human problems which arise from one's social situation, interaction with others, conflicts of desires and values, limits of self—knowledge, practical difficulties (such as poverty and overwork), and so forth. For these the medical approach is inappropriate, and the kinds of help to be given have in common that they should respect and encourage the human potential to take some control, in so far as it is possible, over one's own destiny. It is very important to distinguish the two, even though many people may need, and know they need, a combination of both types of assistance, since even here the different aspects of their problem need different treatment; this is especially important because of the harm that may be done by tackling things the wrong way. Exactly how one formulates in words a practical criterion for distinguishing the two situations is a problem still with us; it is perhaps easier in practice for the therapist, who may be well able to make a correct assessment even if they cannot say why, but it remains as a problem.

The crucial points are, though, the limitations of medicine, and the fact that medicine should in a sense be subordinate to human autonomy: a patient is not a machine to be put right, but a person whose ability to act purposively and successfully is currently interfered with by an illness, disease or malfunctioning, the removal of which should increase their power of decision and action. And this brings us to the general point of this paper: that the improvement and extension of medicine, which have brought great benefits and, one hopes, will continue to do so, should not be allowed to obscure the fact that medicine has limitations in principle as well as in practice. On the one hand, it can never deliver us from our general liability to be physically limited and to suffer, grown old and die. On the other, it cannot form a substitute for our having to decide what to do and work out solutions for our problems in living, and having to work them out not as pure egos but as social and physical beings: a theme developed in other papers in this volume, notably those by David Braine. In between, it needs to be limited by an awareness, first, that sometimes non—interference is the best medical course of action, and secondly, that what is medically best, from the point of view of 'cure', may so adversely affect a patient's autonomy, dignity or even comfort as not to be at all worthwhile: this awareness needs to be shared by doctors and patients alike. In short, if we are to maximise the benefits of medical technology and minimize its harm, there is a need not merely to improve the technology and to be aware of what in practice can and cannot be done, but also to understand the limitations built into the technological approach as such. These limitations are not simply ad hoc, temporary and practical; but arise also from the most basic conditions of human existence, both those relating to our 'animality' and those relating to our ability, and necessity, to be part of a social group, to decide what to do, and to have values and purposes. There is, in other words, a practical need to have a philosophical approach to technology and to see it in general — and medical technology in particular — in a wider human context. This paper is an attempt to make a small contribution towards this.

References

R.D. Laing, *The Divided Self*, Penguin books, Harmondsworth, 1965.
E.K. Ledermann, *Mental Health and Human Conscience*, Avebury, Amersham, 1984.
J. Miller, *States of Mind*, BBC, London, 1983.
D. Smail, *Illusion and Reality*, J.M. Dent & Sons, London, 1984.
T. Szasz, *Ideology and Insanity*, Penguin books, Harmondsworth, 1974.

5 Human life: its secular sacrosanctness

DAVID BRAINE

I Introduction

Human beings are animals; this is intrinsic to their nature and intrinsic to any proper grasp of the shape and basis of human morality. There is a common approach, rooted in a dualism which thinks of a human being as a mind or brain administering its body as a machine or instrument, which undermines this perception of man's animal nature. There is a constant attempt to root the value of human beings in the mental, in isolation from the biological, aspects of their nature, a constant attempt to represent the personal as something that can be segregated from the biological, the bodily and, in short, the animal aspects of human nature and action.

This is not possible even in regard to the issue of killing, the discussion of the right to life and the wrong in killing or in killing the innocent. For, even here, it is the taking of biological life that we are concerned with, because whether or not some life persists after death is a metaphysical or religious question, and it is the termination of bodily human life that is our direct concern. So we need to consider the good in human biological life as such, and, if such life is more sacrosanct than any other biological life, why it should be so; and, if it be significant in particular cases whether this life is innocent or not, we need to consider on what basis being innocent could be considered to make a difference to the value or sacrosanctness of the life in question.

These things we need to consider without depending upon or even drawing upon any premises which involve elevating Nature to divinity or which draw on the command of God or values set by God upon the human life as premises.

Life is in some metaphorical sense given by Nature. When this is said, our attention is drawn to the roots from which human beings spring: we are bodily beings and animals, with dependence on and community with the rest of Nature, both in our origins and in our continued life. But if this saying that life is given by Nature is to have some point in moral discussion we have the presupposition that what is given is a good.

In a parallel way, the theologian may tell us that life is a gift, not of Nature, but of God. In saying this he will have in mind that human life is such as to require not merely divine permission but positive divine activity in giving and sustaining it in existence. Things whose existence consists solely in the absence of other things, e.g. ignorance and holes, without positive existence, need no positive divine activity directly to sustain them in existence. By contrast, life is something positive, requires God's sustaining action, and since God cannot do evil, life must be good, so that to deny its value is a species of blasphemy. Such argument, however, only establishes that if the theological premises are true, then human life must have some inherent goodness. This inherent goodness of human life was what was presupposed in thinking of what Nature gave in giving life as a good. But there must be some source of insight into this inherent goodness in human life, and the special goodness attaching to human as distinct from other biological life and innocent human life as opposed to other human life, which is discernible in principle without appeal to theological premises.

Yet, in regard to all fundamental propositions of value and in morals, there is no demonstrative proof from prior premises. Accordingly, their truth, if not at once perceived, can only be exhibited or shown by rather indirect forms of argument, e.g. the portrayal of a background and the establishment of a perspective within the context of which the rejection of the suggested insight or perceptions appears absurd.

What I say will show that the only life which is worth living for man, the good life, involves a certain wholeness, the integrity or unbrokenness of a certain *Gestalt*, and in virtue of this is also the life which is alone honourable, honest and truthful and therefore worthy for human beings.

The virtues, on this account, are not just habits of acting and feeling in certain ways, which could be defined according to the rules of action and emotion which they incorporate, because thus considered the inner unity which brings together what would otherwise be a mere assortment of rules is lost sight of.

Once we adopt, even if at first only in connection with some difficult and exceptional case, the calculative approach of considering and weighing up the outcomes of various possible courses of action, so as to produce an apparent reason for varying the rule of action indicated by the virtue concerned in order to avoid unhappy consequences in this difficult case, we are in a setting in which whenever dispute arises and it is alleged that there is a real ethical dilemma we have no ground for not resorting to the same calculative method of weighing consequences, and measuring them in terms of pleasure or satisfaction and pain and disappointment, so that the consequentialist or utilitarian approach comes to infect not only our consideration of a small number of difficult cases but to shape our whole consideration of the good life and of virtue for man. It is this utilitarian or consequentialist approach, which measures everything in terms of outcomes, and considers them, as it were, atomistically, as bits and pieces of happiness or unhappiness, upon which modern man falls back when in doubt, and which holds the field for lack of a better or more convincing rounded alternative. Refutation of utilitarianism by indicating the absurdities and insoluble problems to which it gives rise, while adequate and indeed absolutely conclusive at the theoretical level, at the practical level is simply impotent in carrying conviction, broken—backed from the start. What is needed is a portrayal of a whole rival *Gestalt*, a background within which a different perspective sets all problems being considered in a different light, so that, even in those cases where the practical conclusion suggested by the two approaches is the same, the conclusion is rooted in a rounded consideration of human nature and human good as a whole, human happiness being a structured unity.

The rules of action and emotion associated with a fundamental virtue are therefore set more simply by respect for such a general end or *Gestalt*, not calculatively according to a consideration of outcomes or consequences, scattered bits and pieces of upshot atomistically conceived, and therefore variable in detail according to one's assessment of such contingent outcomes.

Accordingly in this paper I wish to consider the virtues [1] in human beings, as self—reflecting animals, which are violated when a proper awe, respect, and value are not set upon human life. I have in mind (i) the virtues of hope and fortitude; (ii) the virtues of humility or truth ('realism' today suggests an unproven pessimism) as the mean between arrogance and poor—spiritedness, and as involving a truthful recognition of both the largeness and the limitations of one's own nature as well as a proper sense of community with one's fellow human beings and with the rest of Nature; therefore also (iii) of the virtues of justice and *pietas*, of which this sense of community is a presupposition.

The possibility of an approach along these lines, and indeed the possibility of any non—atomistic, non—utilitarian, non—consequentialist, approach is undermined by a certain kind of individualism. My first task is to identify this kind of individualism, the task of section II. My next task, in section III, will be to open out the possibility of a basis for human morality, which, by incorporating the biological, entirely escapes this individualism, while doing full justice to human personhood.

II   Individualism: A Systematic Distortion Vitiating Modern Approaches to Questions of Value and Morality

My purpose at this point is only to portray a certain perspective from which most modern approaches to ultimate questions of morals can be viewed. According to this perspective, most of these approaches are ultimately shaped by the assumption that any rationale of human morality must presume a certain individualism. I do not intend to develop this thesis in detail, or to go in for an extended refutation of the systems of moral philosophy which I am attacking: that is a task for a much more extended discussion than this paper requires. Rather, what we need to do is to identify the tendencies in moral philosophy that I label with this word "individualism", in order to point up the distinctiveness of the account of morality which I give in later sections.

From this perspective, then, the individualism which I am picturing is that which regards the questions, "What is the good of it for me?", as the starting point in any deliberation.

Given this approach, one must take it that human beings are not *social* by nature but only *sociable* (capable of society), so that morality is somehow voluntary, depending on the human will and, if human beings are resolved to be consistent, on their consistency.

Granted this context, in order to rescue some remnants of altruism, most often philosophers and political theorists have resorted to the notion of contract, envisaging that it is in the interest of the individual either to make some social contract with others or (to generalise) *to act in ways implying that they consent to live as if they have entered into such contract*. The most recent variant of this approach has been offered by John Rawls in his theory that justice consists in acting according to rules which one would agree upon if sitting round the table with others in a state of ignorance as to one's future circumstances and, from this position, agreeing upon the system of laws which would be safest for oneself.

Alternatively, it has been thought that there is some good for the individual in being consistent and non—arbitrary or impartial, not only in theoretical reasoning, but also in decision and action. But that this good of "rationality" (in this controvertible, specialised, sense) for the individual should take priority over other goods is given little or no justification.

Two tendencies are prevalent, one to assess the good in terms of individual wants or interests, and the other to assess it in terms of impartiality or in terms of its intrinsically general character, what moral philosophers sometimes call "universalisability", whereby all persons in like situations have to be treated alike. These tendencies are sometimes combined; however they have never destroyed human insight into, or inkling of, other inherent goods of a very general kind. Thus insight into the value of such things as knowledge and truth, liberty, the avoidance of false dealing (e.g. in framing or punishing the innocent) has never been lost: the attempt to reduce all other goods to 'satisfaction' and/or 'rationality', and to represent the good in these other things as derivative from their relation to the good in getting what one wants ('satisfaction'), or the good of impartiality, has never seemed attractive, and has generated many very contorted treatments of difficult cases, as well as leaving the distinction between, on the one hand, generosity or supererogation and, on the other hand, obligation obscure, and (perhaps above all) leaving the special significance of family relations ill—explained.

However, even the question "What is the good of it for me?", can be taken in a non—egoistic sense; a man could consider, even if he were dead, that it would be a good for *him* that his children would still be alive, that the human race were not destroyed, that the wonderful system, represented in the actual existence of even a single animal body, or in the existence of the starry heavens, still stood. However, this is not the way in which the question is commonly taken and, neither in the more common interpretation of this question, nor in the Kantian—type requirement of universalisability, do we find any privileged candidate for being counted as a basic paradigm for rationality in practical deliberation. No reason has or can be given for preferring the one to the other or for preferring either to certain others as a paradigm for what should count as rational.

People are often tempted to look for a moral system which they think they can recommend easily to the world. Accordingly, systems of morals in which the motivation of self—interest plays a basic role or within which the perspective of self—interest is widened but only in order to make allowance for what is needed for adults to live together in moderate co—operation, order, amity and safety, or widened yet further to allow of the imaginative consideration that one might have been in the predicament of this or that other human being oneself, are therefore attractive. But what is good and desirable is not the same as what one can easily persuade people to agree to be good or desirable. What we are looking for is a valid basis for a view as to what *is* good and desirable, not a political assessment of what system of values people may be most easily persuaded to agree upon.

Such then is the picture of tendencies in modern moral philosophy, in much of the common thinking of ordinary men and women in western society, and in much political thinking, which I wish to present in sketch in order to make it abundantly plain how another, from this perspective radically different, approach to human morality, is possible.

III   The Basis of Human Morality

If we consider human nature, we find that the context in which human beings take their origin is typically the family, with father and mother and normally with siblings. And this is the context within which some of the goods and obligations which seem most evidently valuable, not only as means, but also as ends worthwhile for their own sake, are to be found. We are in the context of the virtue which the Romans and mediaevals called *pietas*. The concept of *pietas* is most easily understood by contrast with the virtue of gratitude. We give thanks or express gratitude for things done by others for us which are of kinds of which it is, generically at least, in our power to make equal or commensurate return. But to our parents who gave us life and typically nurtured and educated us, we have no power to make commensurate return. And likewise to everything that belongs to our roots, the rest of the extended family, our whole environment, with the Earth, the sea, the clouds and rain, the sun, sky and atmosphere, vegetation, etc., we owe the respect of *pietas*. All these things and our relations and obligations in regard to them, are, as it were, given in the sense of being *data* of our existence, rather than voluntary, in the way in which individual friendships and contracts are normally voluntary.

I have begun my own positive approach with a consideration of kinship and of the virtue of *pietas* associated with it.

The human being is a self—reflective animal. If he considers the essential elements in family relationships he will observe that parents by their genes give shape to their children, and that they are also efficient causes or productive agents in respect of the children, and that they also provide the initial material from which the child grows up. In addition, the mother's womb is the initial context of the child's growth, and the household or home is typically the context of the child's nurture and much of its education. If, then, these provide a reason, or basis in circumstance, for the value as a good, along with the obligations associated with them, within family relations, then man as a self—reflective animal must conclude that the existence of analogous relations between him and the rest of Nature and with the community of human beings rooted in Nature (somewhat as his siblings are rooted with him in their parents) carry with them analogous values and obligations.

Accordingly the nature of man and the character of his origin leave no room for the individualism associated either with egoism or contract theory.

Likewise also they leave no reason for regarding, in a Kantian—style, all human relations as on a par (as if it were only limitation in human powers, spatial distance, and differences in number of occasions of meeting, that gave us greater obligations to those close in family to us).

This escape from an individualistic basis for morals is of key importance in regard to consideration of the right to life. Kantian types of approach (though not the approach of Kant himself) permit the admission to the status of universal laws of action laws whose statement is exceedingly complex, making allowance for vast numbers of varieties of variation in circumstance, and in this way can let past all the exceptions to simple principles such as "You shall not kill the innocent" that the utilitarian might want. The egoistic approach, even when mitigated by considerations of prudence (in earlier tradition, what antiquely was called prudence of the flesh, rather than prudence proper, or wisdom), of entering contracts for mutual self—protection or mutual advantage, or of living as if one had consented implicitly to living under such contracts and consented to developing attitudes appropriate to having thus consented, or of living in accord with principles which (if one were in ignorance of one's possible situation) one would regard as a most safe or advantageous (Rawls), leaves many matters ill—explained: e.g. it leaves obscure the position of the 'others', e.g. those other than myself or who, perhaps for reasons of immaturity, are not amongst those contracting or consenting or not among those who are envisaged as, conjointly with oneself, assessing safety precautions or provisions for mutual advantage for themselves.

It also leaves, in all its versions and adaptations, the basis for assuming the equality of all members of the human species obscure. Once one has turned a blind eye, in one's initial setting up of the basis of morals, to our roots in the community of Nature and the Universe, and begun instead by viewing human beings as deliberating minds, as it were (in our moral autonomy) virtually gods, administering their bodies as they administer machines, one has left no room for re—introducing in any sacrosanct way any principle designed to guard anything into whose definition a biological element is indispensable, whether it be life, the avoidance of pain, the avoidance of bodily harm and mutilation, or anything involving sexuality. If one introduces principles from some tradition, e.g. that the avoidance of the infliction of pain or harm on others has priority over the gaining of benefit for oneself or others, then these principles will turn out to be intermediate ones whose application, if it be in any instance whatsoever questioned, will have to be justified independently, without resort to them, by reference to other supposedly more fundamental principles, attributed to intuition, and commonly utilitarian in kind. (For such intermediate principles, the launching of arguments of egoistic, quasi—contractarian, or quasi—Kantian kinds, may be attempted.)

IV   Truth and Realism as the Basis of Morality

The presuppositions underlying what one judges to be virtues in the human being will be whatever one judges to be a truthful recognition of human nature, roots and general situation. In this paper I have already posed the view that we are animals, albeit self—reflective animals. Accordingly our nature, like that of the other higher animals, is to grow through immaturity within the womb of a mother, through a further extended period of immaturity and development outside the womb within the context of some kind of family, to maturity and adulthood, itself passing through phases of greater physical vigour and less wisdom and experience, to stages of declining physical powers and increased judgment, passing into an old age marked by feebleness in an increasing number of respects and ultimately marked by death. This is the generic pattern which can be departed from in many ways due to early disease, or other disorder, accident, death by violence, etc.

To the same nature it belongs that the human being has roots in a community; proximately, the community of human beings, and, underlying this, the community constituted by Nature as a whole. We cannot on this basis, independent of any appeal to some revelation of the purposes of God or Nature, lay any claim to a right to exemption from pain, disease, death, natural accident, social disorder or other incidents characteristic of the life of social animals. Nonetheless what life we have in its positive aspects is worth having, and its worth is not removed by any of the negative features I have just instanced. The negative features are, as it were, parasitic upon their positively existing subject, the human being, possessed of human nature, characterised by a range of potentialities, some of which are always realised so long as there is life. This subject constitutes the reality presupposed by these negative features. The latter rather represent the absence of some *desideratum*, some development, some serenity, or some other good, of which we may have hope, but to which we cannot lay claim as of right. Some of these negative features, like distress at evil, sorrow for offence, grief at loss, fear of harm, pain connected with damage, and such like, are in different ways integral to being fully human: they and others are also means of growing in humanity, *virtus*, fortitude, hope, wisdom, generosity, etc., as well as occasions of humiliation and of degeneration.

This then is our situation, a situation full of the unknown. When some of our faculties are impaired or lost it remains obscure to us how far the development of other faculties may proceed in a compensatory way, and also obscure to us how far we may hope even for a restoration, whole or partial, of the faculty concerned, whether by natural means through the development of medical or related sciences, or through supernatural means (it is not part of our situation to know that the supernatural does not exist). It is also unknown to us what personality we have or may develop in the face of difficulties, and what effect for inspiration, opportunity for service, or darkness of spirit, we may have on others. Nor do we know to the full even what it is that God or Nature has given us in giving us life, so if we despair of or fail to respect this life in ourselves or in others, we turn aside from something whose full value and full range of potentialities is outside our perception.

It is in this setting that the first two of the virtues I named, namely the virtue of hope and the virtue of fortitude, courage and perseverance, need to be considered. It is a need, for the sake of success, in any long endeavour, a condition of stability, maximum serenity, of keenness of desire and happiness in pursuing such endeavour, that a human being proceed according to both these virtues. To vacillate in respect of either is not only to suffer disadvantage and to subject oneself to additional dissatisfaction or unhappiness, but also to abandon that approach to human life and living which is truthful, realistic, appropriate, for human beings, granted that their nature and situation is as I have described − not a situation for Stoicism, consisting in unmovedness by fear and control of desire, but rather ever a situation for hope. That such abandonment is in this way contrary to virtue is exhibited in its being what Aristotle would have spoken of as poor−spirited, ignoble and dishonourable.

In a certain sense each one of us is a standard−bearer for the rest of the human community, and even, as the one self−reflective part of the animal kingdom on Earth, standard−bearers or representatives for the whole animal kingdom. So in a certain way we let down and dishonour our fellow human beings and the whole animal kingdom, all of which are involved in some manner in strife just as we are, if we despair, or if we do not persevere, even to the end. We are in addition, inasmuch as what human life is in its dimensions is a deeper unknown than the life of other animals, guilty of folly: what we abandon may have in some way infinite or transcendent value − as I have said, we are dealing with something, the value of faithfulness to which, and whose full range of potentialities, are unknown. If what I say is true, then the abandonment of hope or the turning back for lack of fortitude are also violations of justice to our equals and of *pietas*, as what is due out of respect towards our roots, our parents, our home, our environment of fellow beings.

I spoke also of humility or truth as the mean between arrogance and poor−spiritedness. The willingness to take our own life or have it taken can be the expression of poor−spiritedness in a way I have explained through its connection with the abandonment of hope or lack of fortitude. But it can also be the expression of a certain arrogance whereby we insist upon retaining life upon our own terms and conditions and upon no other. We will live in the strength of youth or health or not at all. We will live with this or that personal relationship in the state we desire or not at all. We will live held in honour by others or not at all. In these and in a multiplicity of other ways we judge ourselves driven to the wall, brought to a terminal point beyond which we can stand no more, whereas these judgements are arbitrary, are our own deemings, without basis in fact: we are not at the end; we may yet be driven further; or we may have possibilities within us which we do not realise.

The arrogance exercised in taking our own life may be exercised also in the presumptuous act of asking another to take our life for us, asking him or her to arrogate to himself or herself the right thus to co−operate. But this arrogance may extend further, and, in human history, has extended vastly further, in the direction of taking upon ourselves the right to take the lives of others uninvited − whether in hate or revenge, or for the sake of power, or for some other end.

Because we are self−reflective animals, we realise in our thought that we are social beings, capable of harming other human beings. Therefore it has been held that when a person wilfully co−operates in the harming of others or has so done, or if, even when ignorant or immature, they attack the life of others, then the absolute bar to taking their life may have been removed: thus either (only by civil authority) in war or punishment, or in self—defence (the last case mentioned), if there is no other recourse, then their life may be taken in certain circumstances. But it should be noticed that this qualification, if made at all, is only to be justified in terms of our social nature, which is only integral to us if our animal character is integral to us, and that each of the three exceptions just made, besides being very limited in extent, has been questioned both in detail and *in toto*. I say that these concessions are limited in extent, because it is evident that in most human wars the traditional conditions set for the waging of a just war have not been met, and that these considerations offer not even the beginnings of a justification for taking the lives of non—combatants in a war as a chosen means, or, if when not aimed at, even a foreseen consequence when it dwarfs the tactical end or even parallels it. Nor do they justify the envisaging of the death penalty as a proper punishment for light crimes, or in states of society where (e.g.) prison is a viable alternative (in some societies the powerful would readily kill the garrison and rescue their relatives if the latter were imprisoned). It is evident that in most human wars power, property and prestige have been given a greater value than human life; and in the over—ready willingness to utilise capital punishment some survival of the motivations involved in revenge and some tendency to give priority to respect for authority and property over respect for human life have been almost universal. The over—ready tendency to resort to war and, in the smaller scale matter, this over—ready tendency to resort to capital punishment, both exhibit the arrogance in respect to life of which I have earlier spoken. Yet the key point to be insisted upon is that there would be no room even for these concessions (if there be room even for these), if our animal nature did not give basis for the supposition that we are social by nature, not by consent: it can only be because of this that we might be capable by certain of our actions of removing the absolute bar to taking our lives. Nothing endemic to us by our nature, general condition, state or situation, such as immaturity, handicap, injury, weakness, age, resultant unemployability within the certain social structures ... could remove the absolute bar to the taking of human life without the virtues I have spoken of, viz. a proper *pietas*, humility, fortitude and hope, being violated in folly, arrogance and impiety, wherein, for some short—lived end, the bequest of nature in its unknown dimensions is rejected in oneself or destroyed in others.

In our technological approach to physical nature and to our whole vertebrate environment, and in this most recently in the transformation in agricultural method in "Livestock management", we have already abundantly displayed our capacity for impiety or disrespect for our heritage. But we have now turned, in addition, as it were to savage our own nest, first, in an outburst of Hegelian and Nietzschean *hubris*, and now under the guise of liberalism and compassion, compromised by consideration of criteria of economic functionality and resource scarcity, but all the while preferring the values, education, and necessities of consumerism to any wider perception vouchsafed to human beings.

Meantime, all this continues in a setting of a supreme folly, impiety and betrayal, in respect of both man and nature, wherein a game of risk is played with life on this planet itself. Bertrand Russell did well in his last article for the London 'Times' to speak of the 'cosmic impiety' in man's 'spawning his evil' outside the bounds of the Earth. This was represented for Russell in the appearance that the arrival of man on the Moon was not primarily the product of a noble desire for knowledge or the spirit of adventure, but rather merely the extreme in abuse of resources in an ignobly pursued competition to preserve national prestige and military advantage.

The title of this paper, the 'secular' sacrosanctness of human life, is carefully chosen. To modern man, the word 'secular' normally signifies a self — sufficient in — turnedness of individual men and women and communities on themselves. In origin, the word 'secular' had no such connotation, but suggested a system of spheres or circles, not necessarily self — enclosed or excluding external influence, much larger than humanity, within which we figured as but one inhabitant in the least dignified region of the system, dependent for our origin and continued existence upon the movement, heat and light emanating from higher spheres. We cannot today return to an identical Ptolemaism or geocentrism, but we can retain the use of the word 'secular' and the sense of piety towards the context of our existence and the roots associated with it.

Notes

[1] By virtue, I mean a good character trait whereby we feel emotionally and act appropriately in a setting of diverse emotions and goals of kinds liable sometimes to disturb deliberation and decision, either by disturbing the ends sought after, or by disturbing the discrimination between the different means which may suggest themselves in deliberation.

6 The sanctity of life and the sanctity of death

JOHN HOSTLER

I Introduction

Is it not astonishing to recall how little the medical profession could achieve a century ago? It could cure some common diseases, relieve pain to a limited extent, and perform straightforward surgery; but it was still a chancy business, usually aiming to alleviate the worst symptoms of processes which it could not hope to control. But nowadays it can often claim to be able to control them too. With its modern armoury of drugs, with its sophisticated equipment and elaborate techniques, it can now save, enhance and prolong life to an extent that seems almost godlike.

Unfortunately its astounding technical progress has not been matched by equal advances in moral wisdom and sensitivity. The medical profession generally still appears to hold a rather simple—minded hedonistic philosophy of life and death: it tends to assess the value of life solely in terms of pleasure and the absence of pain, and therefore to assume that the consciousness of death must always be an evil. In consequence patients are sometimes treated as less than autonomous human beings, and sometimes they are unwittingly denied sympathetic care.

This paper explains how and why such mistakes and abuses can occur. It also puts forward a more adequate account of the value of life and death, and explores the implications in the care of people who are terminally ill.

II The Sanctity of Life

Although 'the sanctity of life' is often invoked in debates about medical ethics, its meaning and force are often rather vague. Usually it is meant to condemn the killing of human beings as somehow intrinsically wrong, rather than as wrong merely because of the consequences.[1] But the extent of this condemnation varies: for example, some people contend that deliberate killing of humans is always wrong, no matter what the circumstances, whereas others allow it between armed combatants in war. The grounds for the condemnation vary too. Traditionally, they have been theological: people believe that life is literally 'sacred' because it is created by God, and accordingly they regard killing as a dreadful sin, since it amounts to the destruction or rejection of a divine gift; they consider murder and suicide to be equally wicked since in both cases someone takes a life which is not strictly *his* to take: "he is not the independent lord of his life, but a steward, subject to the sovereignty of God."[2]

But this argument tends to prove either too little or too much. Too much, simply because God has given life to all other creatures besides ourselves: their existence, therefore, should logically be counted as sacred as ours, and there should be no difference in principle (as opposed to in degree) between swatting a fly and murdering a human being. The Buddha and Albert Schweitzer both drew this conclusion, of course, and accordingly they taught their disciples never to harm any form of life intentionally. Most Christian theologians try to evade that universal prohibition by insisting that human and animal life are essentially different, Man alone being made 'in the image of God'.[3] But that is an assertion rather than an argument; and if it is explained (as it is by some Catholics [4]) through the doctrine that only human beings have immortal souls, it just invites further objections. One is that the concept of 'a soul' is notoriously obscure, at least from a purely philosophical point of view; another is that it tends to weaken the force of an appeal to 'the sanctity of life', which may thus prove too little. For if a person's soul is truly immortal, and if it is somehow the real core of their being, it seems that you do not really *harm* them by killing them: all you do, in effect, is to free them from a physical body (which in any case may be counted so much dross) and speed their passage to another and probably a better world. This argument is actually advanced in the Bhagavad—Gita;[5] and though Christians will not accept all its theological premises, they can scarcely deny the substance of its conclusion, that killing may be more a wrong to God than an injury to Man.

Theological arguments therefore supply only a weak defence for the sanctity of life. Even if one has no intellectual objections to belief in God and an immortal soul, the use of those beliefs to prohibit the killing of human beings is beset with intractable logical problems. Consequently secular arguments are to be preferred. Most of those commonly deployed today are 'transcendental' in the Kantian sense: i.e., they prove that life must be valued by showing that it is a necessary condition of something else which is valuable. For instance, if you find certain experiences or activities worthwhile, *a fortiori* you must value being alive, since otherwise you could not be aware of them.

This basic form of argument can be filled out in various ways. Many exponents give it an exclusively hedonistic content, presumably believing that the only experiences which really deserve to be valued are those which are enjoyable or nice. But this version has a worrying corollary: if life is valuable for the sake of pleasure alone, it must be valueless if it becomes irredeemably unpleasant. Some of the later Epicureans actually reasoned thus to justify killing themselves when life ceased to be enjoyable.[6] More recently, and more disturbingly, their argument has been employed to justify killing others too. For instance, it has been used to recommend euthanasia for patients who are incurably ill and who can expect nothing but continuing pain, or to urge infanticide for malformed babies who face a lifetime of suffering: thus Engelhardt tries to prove that there is 'a duty not to prolong a life which has a real negative value for the person involved', where by 'negative value' (as he later makes clear) he just means that it is unpleasant.[7]

Evidently this has become a weak argument against killing people. But then it never was a strong one: hedonism is inherently a poor foundation for the sanctity of life, simply because pleasure is merely one small element in life: a type of experience which is so occasional and so transient cannot reasonably endow the whole of life with value. Luckily there are many other kinds of experience and activity which, though not always wholly nice, are nevertheless valuable: physical exercise, for example, for the health, strength and agility which it develops; learning, for the endless discoveries it affords; relationships with other people, not only for comfort and companionship but also for insights into human nature; and, as an ingredient in all of these, hardship, pain and even suffering, for the self — reliance, maturity and understanding which they may engender. And these are only *some* of the experiences we prize or deem important. We value them, and for their sake we must value life too: "they make life worth living", as we say, and therefore they also make killing wrong.[8]

One advantage of couching the argument in these broader terms is that we see more clearly exactly what it is that we should value. It is not 'life' in Aristotle's 'nutritive' sense,[9] mere biological existence; rather, it is sentient existence: life endowed with a capacity for acting, experiencing and valuing. In this form the argument accordingly establishes that only *human* life is sacred, and it is not in danger of 'proving too much' in the way sketched earlier. But it may still prove too little. Because it presupposes that people are able to have valuable experiences, it does not apply to individuals who lack that ability: thus it does not safeguard the lives of those who cannot experience anything at all, or of those who cannot value their experiences — those who are irreversibly comatose, for example, or those who are still in the womb; perhaps also those whose mental capacities are severely limited.

These are very serious omissions, of course, but I shall not attempt to rectify them here. There is space only to observe that they escape *any* argument of this form: for if you base the sanctity of life upon possessing human characteristics you are bound more or less to omit all those who do not possess such characteristics to the same degree as the rest of us. I have argued elsewhere that the only way to include those unfortunate individuals is by employing some version of the 'slippery slope' argument [10]: for example, Engelhardt contends that the concept of 'a person' must be defended at all costs, since it is central to our whole scheme of values, and therefore he recommends that we should treat *as* persons even those individuals who exhibit only some of the characteristics of personality.[11]

But it would be too much of a distraction to consider that line of reasoning further here, and in any case the purpose of the first section of this paper is now accomplished. It has merely sought to point out that we could not have any valuable experiences if we were not alive, and that therefore life is to be valued. I now want to show that we would not have such experiences if we were not also doomed to die, so that death deserves to be valued too.

III   The Sanctity of Death

My argument for the sanctity of death will be analogous to that for the sanctity of life, but not exactly parallel to it. One obvious source of asymmetry is that although we must be alive in order to have worthwhile experiences, we do not actually need to die. This is shown by the fact that the Gods of Greek mythology pursued ambitions, had love affairs and even went to parties, just like us: as Plato complained,[12] their lives were all too human, save that they neither aged nor died. Subjectively, indeed, many of us are as immortal as they, in that we make our plans and live our lives without any thought of dying — as if assuming that it will never happen to us. In childhood we cannot even understand what dying means,[13] and for a long time afterwards we scarcely think about it: it is said that we do not really begin to see ourselves as mortal until about the age of forty.[14] Up to then, and perhaps even thereafter, our awareness of our fate is 'notional' rather than 'real', in Newman's terminology [15]: we know that 'we must die' as an abstract general truth, but not as an event envisaged concretely, in all its inevitability and with all its implications.

This is evidenced by the fact that many of us never seem to get beyond the thought that we must die 'one day'. That is perfectly true, of course, but it is only part of the truth; the other and perhaps more important part is that we *may* die *any* day. A sudden illness, an accident or an assault may kill us tomorrow — or even in the next few minutes. To assume that 'being mortal' means no more than having to pass away in old age is a classic instance of 'bad faith', as Sartre insists.[16] For death is not just a distant terminus where we all have to arrive at some remote future date; it is also a possibility which is present with us at every instant: "the possibility of no longer being able to be there", as Heidegger memorably describes it.[17] To know that you are mortal is truly to be aware that you may suddenly cease to exist at any moment.

How should this awesome possibility affect the way you value your experience? The mere fact that you *do* value your experience is clearly quite independent of it, for obviously you can find life worthwhile without thinking of death at all. But some writers have suggested that once you do begin to think of it, only certain ways of living can appear worthwhile. A classic statement of this view is Leo Tolstoy's story, 'The Death of Ivan Illich'. It portrays an affluent county judge who has always accepted unthinkingly the values of provincial society: when he discovers that he is incurably ill and near to death he begins also to perceive the shallowness and emptiness of his bourgeois lifestyle; and thus his final act is to renounce it altogether and to embrace instead a more substantial moral creed. Tolstoy's telling of this tale is so convincing and moving that it may well persuade us (as he probably intended that it should) that we too ought to review our values in face of our mortality. But in fact there is no logical reason for doing so. That you may die at any moment merely implies that your future is uncertain; it does not at all jeopardise your present or your past; and therefore it does not entail that what you have always found to be worthwhile should suddenly cease to be so. Of course, people do change their values suddenly, like Ivan, but not always and inevitably as a result of realising that they are going to die. On the contrary, most of us 'tend to die as we have lived', persisting to the end in the same attitudes and adhering to the same beliefs.[18]

So it is quite conceivable that Ivan might have continued to subscribe to his bourgeois values until he died. Nevertheless, the realisation that his death was near would have had *some* effect upon him: it would surely have made his pursuit of those values more urgent. Realising that only a short time was left to him, he would be bound to try and fill it as far as he could with the sorts of experiences and activities he found most worthwhile, whatever they were. To do otherwise, to waste his last days in procrastination or in pastimes which he himself considered valueless, would be plainly irrational, for it would be to go on behaving as if he still had unlimited time ahead. Observe that there is a logical connection here, not a merely psychological one, and that therefore it is universally applicable: we too should be spurred on by realising that the possibility of death is always present. If we were actually as immortal as we like to think, there would be no reason not to waste our time; simply because we are mortal, idleness is wrong: "only under the urge and pressure of life's transience does it make sense to *use* the passing time", as Frankl says.[19]

Of course there is an important sense in which life is transient anyhow, quite independent of the fact that it ends in death; and some hold that its value is established sufficiently by this alone.[20] For we are never quite the same as we were the day before; none of our experiences can be recreated exactly, nor any of our actions precisely reperformed; all the events in our lives are unrepeatable, and therefore (it is argued) they must be cherished as unique and precious. This argument is patently valid, but it does not have much significance in the present context. For although our acts and experiences cannot themselves be erased from the page of history, their effects usually can. "What's done is done", remarked Lady Macbeth [21], but her apparent truism is only partly true: few deeds are so final as murder, so that most of what is done can be undone, too; almost everything passes away with scarcely a trace in the end, as Shakespeare himself frequently bemoans. Consequently what makes an action seem important is not the mere fact that it has been done, but the fact that there may not be sufficient time to undo it. Awareness of life's brevity, not just of its passage, is what really endows every moment of it with significance; and that is why it was the conscious recognition of mortality that chiefly inspired Horace's famous maxim, *"carpe diem"*[22], and its recent American equivalent, "make today count".[23]

People naturally have different ideas as to how best to spend their time; and I do not want to argue here that certain kinds of experience should be valued by everyone. But some kinds clearly deserve to be given priority, such as those which involve or depend on other people. If you want to see the wonders of the world or to read great literature, the sights and the books will probably be there for you always; people, however, may die at any time, and what you want to do with them should therefore be done without delay. You may not wish to follow Epictetus' advice to murmur "tomorrow you or I may be dead" whenever you are with a friend [24], but some such thought would undoubtedly help you to make the most of the opportunities which your relationship presents.

In ways such as this the consciousness of death must inevitably affect your valuation of experience. If you fully accept your own mortality, you see that your life is fleeting and liable to end at any time, and that therefore it is precious and not to be wasted. That is why I speak of "the sanctity of death". I do not mean that death itself is good, for it brings to an end all experiences you value and therefore it can only be regarded as an evil.[25] But the consciousness of death is good, because that brings home to you the fact that some experiences *are* valuable, and it spurs you to obtain them without delay. If you do not believe that you will die, you have no good reason for getting on with living: "you live your life in preparation for tomorrow or in remembrance of yesterday, and meanwhile each today is lost", warns Kubler—Ross.[26] Ultimately, the knowledge that you are mortal is what makes your life worth *living*, and therefore it deserves to be prized and sanctified as much as life itself.

IV Medical Ethics

So far I have said nothing about the topic of medical ethics. That has been a deliberate omission, for it is surely a cardinal mistake to theorise about medical problems in isolation and to solve them by moral rules which cannot be followed in other areas of conduct. In the first two sections of this paper I have accordingly sought to establish two very general principles: viz., that life on the one hand and awareness of death on the other are both precious and important. Now I want to relate these to the practice of medicine.

An initial point of purchase is the fact that people who work in the medical and paramedical professions publicly uphold the sanctity of life.[27] In so doing they should also acknowledge the sanctity of death: they should accept that full awareness of death helps to make life worthwhile, and therefore they should openly admit to themselves and to their patients both the fact and the implications of being mortal. They could probably introduce such thoughts at many points in their daily work, but the most obvious situation is when they are dealing with people who are terminally ill and who are spending their last weeks or months under medical care. Such patients above all should be helped to understand and to come to terms with the fact that they are about to die.

Unfortunately much current medical practice aims to keep them from thinking about it. Surveys indicate that up to 90% of hospital doctors will refuse to tell a patient that he has an incurable and fatal cancer, despite the fact that more than half of them would want to be told if they had it themselves.[28] So far from telling him, indeed, they will lie outright, saying (almost until he dies) that his condition is not serious; and they will collude with nurses and even with relatives to maintain the deception.[29] They say that they do so in order to sustain his hope and keep his cooperation; but that seems rather a thin excuse in view of the fact that patients who *are* told the truth commonly remain hopeful and cooperative to the very end.[30] So it appears that the doctors' policy is based on prejudice, and that their underlying motives are at least partly irrational. Szasz argues that they have a profound emotional resistance to any acceptance of death, arising from their desire to be professional life—savers for whom death is the ultimate enemy.[31] Freud's philosophy suggests a more general explanation [32]: like Spinoza [33], he maintains that no one can ever accept death fully, at the deepest level of unconscious thought; and in fact, as I mentioned earlier, any people do persist throughout life in thinking of themselves as more or less immortal.

But no matter how the doctors' policy is to be explained, it clearly cannot be excused. Lying and deceit are incontrovertibly immoral, whether they take place in a hospital or not. As Fletcher explains, the most elementary ethical considerations imply that a patient ought to be treated as a moral agent: his personal autonomy and freedom of choice ought to be respected, and therefore he has a fundamental moral right to any information which is necessary for him to make a decision about his own predicament; if he asks for it, he must be told the diagnosis even if it is fatal, since morally "a doctor is obligated to tell the truth to his patient".[34] Now that conclusion is important, so far as it goes; but it does not really go far enough. Consider the reported practice in the cancer ward of one large American hospital where the official policy is always to tell the truth: 'the doctor walks into the patient's room, faces him, says "It's malignant", and walks out.'[35] The doctor has undeniably discharged his strict moral obligation to the patient, at least as Fletcher defines it, but he has not at all helped the patient to *understand* that death is imminent in the sense I am arguing for here. He has merely announced a fact which is so vast and dreadful in its implications that the patient (in all probability) will just not be able to take it in, and will either fail to comprehend it properly or will flatly deny it. Such awesome news surely has to be broken gradually, and the patient has to be led gently to work out what it will mean. As explained in the second part of this paper, when you realise that you are going to die you see that you ought to spend your remaining time in the most worthwhile way you can: accordingly the dying patient has to decide what activities and experiences he values most, and he has to find out how far they are still possible, given the debilitation of his illness, the restrictions of hospital routine and similar limitations. In short, he has to work out the implications of his own impending death in terms of his personal system of values. To help him do this obviously demands far more than the curt announcement of a fatal diagnosis: it requires time, sympathy, tact, and maybe even professional counselling.

Unfortunately this kind of help is not widely available; but then the need for it is not widely recognised. Most writers on the subject of terminal care seem not to accept that a patient should face death in whatever way he finds best, for apparently they do not even perceive that there is more than one way in which it might properly be faced. They subscribe unanimously to the same paradigm of death, agreeing almost without exception that ideally it should be a calm and dignified passage out of life. To cite just one typical example: Hinton, in what is still a standard work on the subject, advises that patients should be given 'spiritual and bodily help' so that they may "experience the peace of surrendering to their fate before they drift into permanent unconsciousness".[36]

This is undeniably an attractive paradigm of death. It is as attractive now as it has ever been, and it has certainly been popular for long. After all, it is essentially the same as the ideal death of the ancient philosophers, demonstrated by Socrates and the Stoics; it is frequently portrayed in Victorian fiction, too, where the usually consumptive hero dies with similar fortitude and resignation.[37] Thus it is no surprise to find Hinton's very words anticipated (rather more beautifully) by Shelley [38]:

"Mild is the slow necessity of death;
The tranquil spirit fails beneath its grasp,
Without a groan, almost without a tear,
Resigned in peace to the necessity,
Calm as a voyager to some distant land
And full of wonder, full of hope as he."

Presumably death *may* be as nice as that. But obviously it is not so always; and simply because of that, calm acceptance cannot be assumed to be the only proper attitude towards it. A more reasonable reaction may be stunned incomprehension, for example: to realise that you *are* going to die must be such an overwhelming and annihilating thought that the most honest response may be to admit that you just cannot understand it. That is why Empson describes death as a topic "that most people should be prepared to be blank upon".[39] Alternatively, like many patients in terminal care, you may feel isolated and forlorn. This too is a reasonable attitude to what Tillich calls "the ultimate loneliness of having to die" [40]: for indeed dying is one thing that no one can do for you, in the doing of which you are utterly sundered from everyone else. Surely isolation and withdrawal are sensible reactions in such a situation? But they are not acceptable in hospital, alas: as Glaser and Strauss report, the dying patient is expected to conform to a model of "courageous and decent behaviour ... (he) should maintain relative composure and cheerfulness ... he should not cut himself off from the world, turning his back upon the living; instead ... if he can, he should participate in the ward social life."[41] A third possible response is anger. This too is very common [42], and again it is surely an understandable reaction to the prospect of being deprived for ever of all that you value most. To accept that prospect calmly, as expert opinion recommends, may well appear to some just a craven and pathetic surrender — as it did to Dylan Thomas, for example, who exhorted his dying father to "rage, rage against the dying of the light."[43] But such defiant fury will also be frowned upon by medical practitioners. Hinton, who must be counted one of the more thoughtful and sensitive writers on this subject, describes people who are unwilling to die as merely "unfortunate", apparently assuming that their reluctance must be just a form of cowardice.[44]

Three centuries ago Jeremy Taylor complained that "it is a sad thing to see our dead go out of our hands; ... the last scene of their life, which should be dressed with all spiritual advantages, is abused by flattery and easy propositions, and let go with carelessness and folly."[45] It seems that in some respects we have not advanced much since then. Certainly, as I have tried to show, if you die in hospital today you will probably encounter a fair number of 'easy propositions'. The very fact that you *are* dying may well be concealed from you for as long as possible, perhaps until you can no longer understand it properly; or you may be informed of it, but expected to accept it quietly and to conform in other ways to an established ideal of deathbed behaviour. You will probably not be treated as an autonomous moral agent, encouraged to choose your own response to the news; nor will you be helped to relate it to your own scheme of values, or to decide how best to spend your remaining days.

Why does this somewhat constricting form of terminal care persist in modern medicine? One reason may be that many writers on the subject tend to describe patients' responses to death as 'stages': thus they imply that there is a standard sequence of attitudes, terminating in quiet acceptance, through which a dying person ought to pass. This model of attitudinal 'progress' is widely accepted and is probably quite influential, even though it rests on very weak foundations. For in fact patients' attitudes do not succeed each other in a regular order [46]: they change unpredictably and often coexist simultaneously, so that it is quite legitimate to see them as a variety of moods which may be manifested from time to time, rather than as an unvarying series appearing in succession. Admittedly patients who survive long enough often end up feeling calm about dying, but that does not imply that resignation is a more reasonable or more mature reaction than any other; for the patients may simply be too tired by then to feel anything else, or they may have been socialised into adopting the attitude most approved of in that environment. Considerable psychological pressure may sometimes be brought to bear on them to make them conform in this way, in order to assist the smooth running of the hospital; for obviously a patient who dies in blissful ignorance or in quiet resignation will be easy to manage, whereas one who is bitter or who struggles will distress the doctors and nurses, as they openly admit.[47] But there are other and deeper reasons besides. The style of dying which the hospital staff find most convenient is also the one which the rest of us find most attractive, as I have already pointed out; and thus it fits in most neatly with the general hedonism that seems to prevail in our society today. We not only claim the pursuit of happiness as a fundamental right but also engage in it as our chief occupation, believing, as Ivan Illich did, that, on the whole, life ought to be pleasant and comfortable. Some say that this leads us to deny the reality of death [48], but surely that is not strictly true: rather, we *distort* its reality, assuming, like Ivan, that death should be comfortable and pleasant too.

But of course it is not; and therefore, when we perceive what it is actually like, we are liable to be intellectually and emotionally overwhelmed — again, just like Ivan. I have been arguing in this final section that people who undertake to care for us when we are dying should help us to cope with that perception of death's true reality: they should help us to understand it and to respond to it in terms of our own values. But I am not suggesting that they will find this an easy task or even that they will be able to achieve it often. There are many practical obstacles, for a start: some dying patients do not have enough time, or are not sufficiently conscious, to comprehend their situation fully, while others lack the courage or the maturity to respond to it positively. And there are theoretical difficulties too. My argument for the sanctity of death does not yield any firm rules about how to care for dying people, for the objective which it prescribes is to help each patient to formulate his *personal* response to death, and that obviously has to be accomplished differently in every case.

Besides, there are other objectives to be considered, which may well conflict with it: keeping patients free from pain, for instance, may be impossible if they are allowed to remain fully aware of their predicament. So 'the sanctity of death' is *not* an overriding principle which magically resolves all ethical dilemmas, any more than 'the sanctity of life' is.[49] It is merely an ideal to be aimed at: an ideal to be weighed against other ideals, and tailored to the possibilities of each person's situation.

Therefore it may be objected that what I have said in this paper does nothing to simplify the ethics of terminal care. That is quite true; I would merely deny that it constitutes an objection. For as soon as medicine begins to deal with *people* (rather than with their bodies alone) it runs into a mass of difficult moral problems: problems which are at least as complex as any which we encounter in everyday life, which usually deal with issues of much greater moment, and which often concern people who are near the limit of their physical and emotional resources. A philosopher does no one any service by inventing some glib formula to solve them, or by suggesting that the solution ought to be easy. His job is surely to analyse and explain their complexity, to show why they are so difficult to solve; and having done that he should shut up, apart from wishing good luck to the doctors and nurses who must actually find a solution.

Notes and References

[1]   This is pointed out by Jonathan Glover: *Causing death and saving lives* (Penguin Books, Harmondsworth 1977) p. 41.

[2]   Bernhard Haring: *Medical ethics* (St Paul Publications, Slough 1972) p. 69.

[3]   E.g.: *On dying well — an Anglican contribution to the debate on euthanasia* (Church Information Office, London 1975) p. 18.

[4]   E.g., C.J. McFadden: *Medical ethics* (F.A. Davis Publishing Co., Philadelphia 1967 — sixth edition) p. 228.

[5]   *Bhagavad—Gita*, chapter 2.

[6]   So claims Raanan Gillon: 'Suicide and voluntary euthanasia — historical perspective', in: *Euthanasia and the right to death*, ed. A.B. Downing (Peter Owen, London 1969) p. 175.

[7]   H.T. Engelhardt: 'Euthanasia and children — the injury of continued existence', *Journal of Pediatrics* vol. 83, no. 1 (1973) p. 170—171.

[8]   This point is argued further by Thomas Nagel: 'Death', in his *Mortal questions* (Cambridge University Press, Cambridge 1979) p. 9—10.

[9]   Aristotle: *De anima*, 415a.

[10]   John Hostler: 'The right to life', *Journal of Medical Ethics* vol. 8, no. 3 (1977) p. 144.

[11]   H.T. Engelhardt: 'Medicine and the concept of person', in *Ethical issues in death and dying*, ed. T.L. Beauchamp and S. Perlin (Prentice—Hall, Engelwood Cliffs 1978).

[12]   Plato: *Republic*, 378ff.

[13]   Myra Bluebond—Langner: 'Meanings of death to children', in *New meanings of death*, ed. H. Feifel (McGraw—Hill, New York 1977).

[14]   John Hick: *Death and eternal life* (Collins, London 1976) p. 88.
[15]   J.H. Newman: *Essay in aid of a grammar of assent*, chapter 4.
[16]   Jean—Paul Sartre: *Being and nothingness*, trans. Barnes (Methuen, London 1969) p. 536.
[17]   Martin Heidegger: *Being and time*, trans. Macquarrie and Robinson (Basil Blackwell, Oxford 1967) p. 294.
[18]   J.D. Harte: 'The care of patients by the family doctor', in *The dying patient*, ed. R.W. Raven (Pitman Medical, Tunbridge Wells 1975) p. 53.
[19]   Viktor Frankl: *Psychotherapy and existentialism* (Penguin Books, Harmondsworth 1973) p. 88.
[20]   This was argued in the discussion when this paper was first read.
[21]   Shakespeare: *Macbeth*, Act 3, scene 2.
[22]   Horace: *Odes*, Book 1, no. 11.
[23]   Orville E. Kelly: 'Make today count', in Feifel, *op. cit.* (n. 13).
[24]   Epictetus: *Moral discourses*, Book III no. 24.
[25]   This is pointed out also by Nagel, *loc. cit.* (n. 8).
[26]   Elizabeth Kubler—Ross: *Death — the final stage of growth* (Prentice—Hall, Engelwood Cliffs 1975) p. 164.
[27]   Cf. the Geneva Convention Code of Medical Ethics, and the International Code of Nursing Ethics, which speak respectively of 'respecting' and 'conserving' human life.
[28]   Donald Oken: 'What to tell cancer patients — a study of medical attitudes', in *Ethical issues in death and dying*, ed. R.F. Weir (Columbia University Press, New York 1977) p. 13, 19.
[29]   B.G. Glaser and A.L. Strauss: *Awareness of dying* (Aldine Publishing Co., Chicago 1965) p. 29ff.
[30]   Oken, *op. cit.* (n. 28), p. 18.
[31]   Thomas L. Szasz: 'The ethics of suicide', in Beauchamp and Perlin: *op. cit.* (n. 11) p. 136.
[32]   Quoted in Raven, *op. cit.* (n. 18), p. 106.
[33]   Spinoza: *Ethics*, Book 3 prop. 10.
[34]   Joseph Fletcher: *Morals and medicine* (Victor Gollancz, London 1955) p. 36.
[35]   Glaser and Strauss, *op. cit.* (n. 29), p. 125.
[36]   John Hinton: *Dying* (Penguin Books, Harmondsworth 1972) p. 108.
[37]   E.g., the death scene in George Eliot's *Amos Barton*, of which the authoress was especially proud.
[38]   Shelley: *The daemon of the world*, part 2, lines 468—473.
[39]   William Empson: 'Ignorance of death' in his *Collected Poems* (Chatto and Windus, London 1956).
[40]   Quoted in Raven, *op. cit.* (n. 18), p. 109.
[41]   Glaser and Strauss, *op. cit.* (n. 29), p. 86.
[42]   Elizabeth Kubler—Ross: *On death and dying* (Tavistock Publications, London 1970), p. 44ff.
[43]   Dylan Thomas: *Collected Poems 1934—1952* (J.M. Dent and Sons, London 1952), p. 116.
[44]   Hinton, *op. cit.* (n. 36) p. 102.
[45]   Jeremy Taylor: *Holy dying*, dedicatory letter.
[46]   This is admitted by Kubler—Ross, *op. cit.* (n. 42), pp. 122, 235.

[47]   E.g., 'Varieties of death on the ward', in *Dilemmas of dying —*
         a study in the ethics of terminal care, ed. Ian Thompson
         (Edinburgh University Press, Edinburgh 1979) p. 19.
[48]   E.g., Philippe Aries: *Western attitudes toward death* ... (Johns
         Hopkins University Press, Baltimore 1974) p. 93—94.
[49]   This is explained further by Daniel L. Callahan: *Abortion — law,*
         choice and morality (Collier—Macmillan, New York 1970) p. 325.

7 Down the slippery slope

DAVID LAMB

I The Slope Argument

The "slippery slope" argument, sometimes referred to as the "thin end of the wedge" argument, can be found in all areas of policy making. Its most common employment is in the form of a rejection of a new proposal in which the argument is advanced not so much as an objection to the proposal *per se*, as to the undesirable position to which it will lead. The slope argument is usually dismissed as an indirect argument which conflates the possible with the actual. Nevertheless, policy makers use it regularly and it is one of the central arguments in the ethics of medical care. Opponents of voluntary termination of pregnancy and euthanasia have employed versions of the slope argument in an appeal to what they predict as the long range consequences of the legalisation of these practices. If, for example, we permit euthanasia for humane reasons, are we not in danger of opening the door to a more callous disregard for human life? What next? The killing of mentally retarded children, those whose maintenance is costly, the unfit, old, or socially undesirable? Would it lead eventually to the killing of the politically and racially undesirable? In this way the parade of Nazi atrocities can be brought out to block off any attempt to introduce the legalisation of accelerated death. Yet, despite some obvious objections to the slope argument, it can be argued that its opponents have not provided an adequate refutation and that, properly understood, there is a place for this argument in ethical decision making. However, its plausibility is greater when utilized in opposition to the killing of the dying than to the voluntary termination of pregnancy.

The most extreme example of a slide into moral depravity is the Nazi euthanasia policy which began with discussions on merciful killing in 1933. By 1936 it was so widely discussed that additional proposals for the elimination of the physically and socially unfit were suggested only incidentally in an article published by an official German medical journal. Hitler signed the first direct order for euthanasia in September 1939. The category for extermination included the mentally defective, psychotics (particularly schizophrenics), epileptics, and people suffering from various infirmities of old age, and various organic neurological disorders, such as infantile paralysis, Parkinsonism, multiple sclerosis, and brain tumours. In the initial stages Jews were excluded from this "privilege" of "merciful release". Charities were utilized to give respectability to this programme. The "Charitable Transport Company for the Sick" delivered patients to extermination centres and the "Charitable Foundation for Institutional Care" collected the cost of extermination from relatives. Over 275,000 people were exterminated in this programme. There was little evidence of resistance, although there were some demands for "legislative regulation providing some orderly method that will ensure especially that the aged feeble—minded are not included in the programme".[1]   These demands, however, can be seen as a particularly notorious way of inserting the thin end of the wedge. Shifting the arena of discussion to the limits of a programme is often a means of establishing its plausibility; and in this case the scope for moral concern was confined to particular cases rather than the programme as a whole.

Leo Alexander, who was prominent in the drafting of the Nuremberg Code, is a forceful exponent of the slope argument. Commenting on the crimes committed by members of the German medical profession during the Third Reich, he said:

> Whatever proportions these crimes finally assumed,
> it became evident to all who investigated them that
> they had started from small beginnings. The beginnings
> at first were merely a subtle shift in emphasis in the
> basic attitude of the physicians. It started with the
> acceptance of the attitude, basic in the euthanasia
> movement, that there is such a thing as a life not
> worthy to be lived. This attitude in its early stages
> concerned itself merely with the severely and chronically
> sick. Gradually the sphere of those to be included in
> this category was enlarged to encompass the socially
> unproductive, the ideologically unwanted, the racially
> unwanted and finally all non—Germans. But it is important
> to realise that the infinitely small wedged—in lever from
> which this entire trend of mind received its impetus was
> the attitude towards the nonrehabitable sick.[2]

Alexander goes on to describe three routes whereby the German people and their medical authorities travelled from euthanasia to the extermination camps. The first route involved the introduction of ethical dilemmas with in—built presuppositions of social utility and in which the alternatives were already rigged.   For example, schoolchildren were set mathematical problems like "how many new housing units could be built and how many marriage—allowance loans could be given to newly—wedded couples for the amount of money it cost the state to care for the crippled, the criminal and the insane?".[3]   Having established the plausibility of extermination on cost—benefit terms, it quickly progressed to those "unfit for work", and then became the nucleus of plans to exterminate Jews and Poles, and to cut down the Russian population by 30 million.

The second route down the slope lay in the brutalization of medical workers. Doctor Hallerworden, a recipient of 500 brains from extermination centres, recalled that:

> The worst thing about the business was that it
> produced a certain brutalisation of nursing
> personnel. They got to simply picking out those
> whom they did not like, and the doctors had so
> many patients that they did not know them, and
> put their names on the list.[1]

The third route lay in the attempt to restrict the concept of medical care to rehabilitation. It is to the credit of Dutch physicians in occupied Holland that they recognized and resisted this step. When they attempted to draw Dutch physicians into the orbit of German medical practices, the Nazis did not ask them to kill their patients. Their request was couched in more acceptable terms. Superficially, the following statement from the Reich Commissioner for the Netherlands is harmless:

> It is the duty of the doctor, through advice and
> effort, conscientiously and to his best ability,
> to assist as helper the person entrusted to his
> care in the maintenance, improvement and
> re—establishment of his vitality, physical
> efficiency and health. The accomplishment of this
> duty is a public task.[4]

The apparent innocence of this statement obscured the fact that it called for the concentration of efforts on rehabilitation of the sick for useful labour, and the abolition of medical confidentiality. This step was resisted despite the arrest of 100 Dutch physicians who were sent to concentration camps. Not a single euthanasia or non—therapeutic sterilization was recommended or participated in by Dutch physicians. Alexander attributes this to a refusal to step onto the slope: "It is the first innocent step away from principle", he says, "that frequently decides a career of crime. Coercion begins in microscopic proportions."[4] These remarks serve as a warning of very grave dangers which follow the acceptance of the belief that only those with an optimistic prognosis should receive the best treatment. For Alexander, the path down the slope is one of logical necessity: "From the attitude of easing patients with chronic diseases away from the doors of the best type of treatment facilities available to the actual dispatching of such patients to killing centres is a long but nevertheless logical step".[4]

Nevertheless, despite resistance by legislators and physicians to lending support to proposals which might be seen as a step down the slope, philosophers have been virtually unanimous in their criticism of the slippery slope argument. In what follows we shall consider objections to the slope argument which have been made by opponents of the sanctity of life argument, advocates of beneficient euthanasia, voluntary euthanasia, cost benefit criteria for life extension, and voluntary termination of pregnancy.

II The Slippery Slope and the Sanctity of Life

According to Alexander, once we accept the concept of a "life that is not worth living" we have stepped on to the slippery slope: any breach of the sanctity of life principle will have disastrous consequences. Marvin Kohl rejects this argument and replies to the slope argument as follows:

> They maintain that if euthanasia is legalised, or
> even held to be moral, then all sorts of disastrous
> consequences will follow. We reply that it is equally
> possible that it may not be abused. Logically speaking
> the point is telling. Unfortunately, however, it is not
> persuasive. Following Joseph Fletcher, we then ask: "What
> is more irresponsible than to hide behind a logical
> possibility that is without antecedent probability?"[5]

According to Kohl, the slope argument rests on an appeal to logically possible consequences, whereby anything we can imagine, without logical contradiction, could be seen as a possible consequence. "Since logical impossibility is an extremely limited kind of constraint", he says, "it neither marks off nor prohibits inconfirmable, false, or conceptual sentences".[5] Hence we can say with equal impunity "that the man may run the mile in two minutes, or that he may not; and that euthanasia may lead to abuse, or that it may not".[5]

The above argument, however, is based on a caricature of the slope argument. Given a specific social context, there may be very plausible grounds for inferring the likelihood of further steps down the slope. Thus, in the case of the Dutch physicians cited above there were very sound reasons for the expectation of further steps, once the meaning of the statement they were asked to sign was made clear to them. Kohl's rebuttal of the slope argument is one which meets a prediction which may be based on highly plausible grounds with an argument designed to refute the fantasies of logical possibility. However, this is not Kohl's central thesis, which he reserves for what he sees as a stronger version of the slope argument. For the defender of the principle governing the sanctity of life, "one ought never to kill an innocent human being because such action *must* lead to undesirable consequences"[6]; namely, that, once embarked on, a programme of killing will gather momentum. Rejecting this argument, Kohl cites "overwhelming evidence indicating that human beings compartmentalise their experience and ideas; and that it is only when the normal process of compartmentalisation breaks down that one encounters difficulties".[7] But here again a context is lacking. If one accidentally, or unwillingly, killed an assailant, it may not follow that one will become a habitual killer. But we can have reasons to fear moral contagion in cases when someone is attempting to overcome his scruples against taking human life. In *The Godfather* the young Michael contemplates only the killing of Captain McClusky as a serious and final act: after the first killing the others are just business.

Kohl's critique of the slope argument and subsequent defence of euthanasia rests on the argument that human beings can limit their generalizations. Often the limit to generalization is the concept of "the same kind" or "same class". "If we crush an insect and believe it is a permissible act we do not believe that it is permissible to kill all living things. We conclude that it is permissible to kill this insect or at most, all kinds of insects".[7] Similarly, "if we are taught to kill Nazis and the criteria for a Nazi and the circumstances of permissible killing are clearly spelled out, we do not kill all German nationals ... We do not mistakenly generalise and kill all Europeans. Nor do we proceed either in fact or in mind to kill all human beings".[8] Hence, "the merciful killing of patients who want to die does not necessarily lead to the killing of the unwanted or the extermination of the human species".[9]

There are two problems with Kohl's appeal to our ability to compartmentalize. First, the problem of "spelling out"; what constitutes membership of a category? (Even with the category of "Nazi" we have problems: do we include all members of the Nazi party or exclude those who may have been coerced to join? And how do we define coercion?) Secondly, it is pertinent to question whether categories are as watertight as Kohl suggests. Even in wartime when the category for rightful killing may include only the enemy, there are genuine fears that soldiers will become desensitized, and efforts are taken to combat this before returning them to civilian life. When we consider conceptual distinctions abstractly, the appeal to compartments has plausibility, but in a social context conceptual distinctions have rough edges. This is precisely what is indicated by the slope argument. Thus Kohl may propose a new boundary out of a morally commendable motive to relieve suffering. But the force of the slope argument lies in the ability to draw attention to the way in which new boundaries can be exploited by those with evil intentions. There is little doubt that the Nazi euthanasia policy was a travesty of merciful killing; that from the start their leaders had thought in terms of a larger scale of killing. But there is evidence that they realized that their programme would have to be introduced gradually, through the extension of newly accepted conceptual boundaries. Alexander records how "the German people were considered by the Nazi leaders more ready to accept the extermination of the sick than those for political reasons. It was for that reason that the first exterminations of the latter group were carried out under the guise of sickness".[10] Thus the conceptual boundaries of psychiatric sickness were redrawn to the point where they included "inveterate German haters", namely prisoners who had been active in the Czech underground.

Whilst Kohl's criticism of the slope argument turns on the problem of drawing up and maintaining conceptual boundaries, Singer's programme for "unsanctifying human life", and consequent defence of euthanasia, rests on the rejection of morally significant boundaries between species. Nevertheless, the appeal to compartmentalism is upheld by means of an assumption of cultural relativism.[11] Singer confronts the slope argument as a "practical argument" for the sanctity of life, and his attempt to overcome it is as follows:

People are liable to say that while the doctrine
may be based on an arbitrary and unjustifiable
distinction between our own species and other
species, this distinction still serves a useful
purpose. Once we abandon the idea, this objection
runs, we have embarked on a slippery slope that may
lead to a loss of respect for the lives of ordinary
people and eventually to an increase in crime or to
the selective killing of racial minorities or
politically undesirables. So the idea of the
sanctity of human life is worth preserving because
the distinction it makes, even if inaccurate at some
points, is close enough to a defensible distinction
to be worth preserving.[12]

Singer's reply to the slope argument here consists of a denial of the
contagiousness of killing by means of an appeal to historic evidence.

Ancient Greeks ... regularly killed or exposed
infants, but appear to have been at least as
scrupulous about taking the lives of their
fellow citizens as medieval Christians and
. modern Americans. In Eskimo societies it was
the custom for a man to kill his elderly parents,
but to murder a normal, healthy adult was virtually
unknown. White colonialists in Australia would
shoot Aborigines for sport, as their descendants
now shoot kangaroos, with no discernible effect on
the seriousness with which the killing of a white
man was regarded. If we can separate such basically
similar beings as Aborigines and Europeans into
distinct moral categories without transferring
our attitude from one group to the other, there
is surely not going to be much difficulty in marking
off severely and irreparably retarded infants from
normal human beings. Moreover, anyone who thinks
that there is a risk of bad consequences if we
abandon the doctrine of the sanctity of human life
must still balance this possibility against the
tangible harm to which the doctrine now gives rise:
harm both to infants whose misery is needlessly
prolonged, and to non−humans whose interests are
ignored.[13]

For Singer, the rejection of the slope argument is dependent upon the
acceptance of an ethical relativism that is hard to comprehend. If ancient
Greeks practised infanticide, white Australians shot Aborigines, and Eskimoes
murdered their parents without overwhelming reasons to justify their actions,
this fact would demand the moral damnation of those Greeks, white
Australians, and Eskimoes who engaged in these activities. Once they have
gone so far down the slope there is little virtue to be derived from the

categories they excluded from murder. But what the appeal to ethical relativism obscures in these cases is the fact that societies are not the homogenous entities they are sometimes made out to be. Not every Greek supported infanticide, nor was parenticide endorsed by Eskimo societies who had the resources to maintain the elderly. In the contemporary world species slaughter is being increasingly opposed. These rather obvious facts are overlooked by both supporters and opponents of the slope argument, whose theses turn on a crude homogenous model of social change, which can be depicted as follows. *Stage I* pictures a society with no support for genocide. *Stage II* pictures one which allows certain forms of merciful killings; whereas *Stage III* offers full genocide. Supporters of the slope argument argue that stage I leads inevitably to stage III, whereas their opponents claim that we can hold the line at either stage. Against both views it is important to point out that societies are not unified in this way, and that within any given society at any given time there will be supporters and opponents of all three stages, plus many more besides. The plausibility of the slope argument is not based so much on the claim that stage I must inevitably lead to stage III, but from the fact that whilst the granting of stage II may serve as a satisfactory position for those seeking no more than stage II, it will nevertheless be seen as a partial victory for those committed to the realization of stage III.

It is necessary to stress this point against the tendency to think that steps from euthanasia to genocide are only taken in a Nazi culture, not in the democratic West. Kohl, for example, sees the Nazi culture as a counter−example to arguments concerning the contagiousness of killing. The Nazis, he says, were acting in accord with a political ideology which rested on the principle that if the proper authorities believed that killing was good for Germany, then it was justifiable. As such he does not see the Nazi case as evidence to support the slope argument. "Rather it is evidence that when men have almost unlimited power, their actions will be consistent with their beliefs, and when their beliefs entail needless cruelty, so will their actions".[14]

This point is not disputed, but it is important to counter the relativist premise on which Kohl's argument rests. The force of the slope argument is not in an appeal to what is logically possible in any society, but rather in the ability to indicate that once on the slope there are, within the Western democracies, very real pressures eager to push in the direction of the Nazi programme. This may indicate a commitment to the principle of pessimism − if the worst can happen it will − but it serves to counter a dangerous naivety with regard to the question "Could it happen here?" To be sure, racial extermination was characteristic of the Nazis' objective from the beginning. No one should think that it all began with innocent proposals to alleviate suffering − although there may have been innocent liberals who were duped into supporting the initial stages of the programme for these reasons. Nonetheless, the following example suggests that we should be wary of claims that the "final solution" has no base of support in the Western democracies.

In 1972, 570 university students in the U.S. were asked to participate in a research project to work out ways of killing "unfit" persons as a "final solution" to problems of overpopulation and personal misery. They were told that "mercy killing" was considered by most experts as not only being beneficial to the unfit, in that it put them out of their misery, but also beneficial to the healthy, fit, and more educated segment of the population. They were told that the only problem was to determine which method of killing should be used, and who should decide when killing should be resorted to. They were told that the findings of the study would be applied to humans once the system had been perfected. Out of the 570 students, 226 approved. When subjects were told that the findings would be applied to minority groups, the acceptance rate was even higher.[15]

Examples of this kind indicate a measure of support for any point on the slippery slope, despite the fact that we do not live in a culture that is outwardly committed to a Nazi ideology. Appeals to the sanctity of life should not be lightly dismissed lest we lend encouragement to those with a total disregard for human life.

III Beneficient Euthanasia

According to Philippa Foot, euthanasia should not be considered unless it is, in some sense, beneficial to the patient.[16]   As long as this principle is maintained, so the argument goes, there is no danger of a slide into a callous disregard for human life. As a defender of beneficient euthanasia, Kohl is obliged to counter the slope argument that this opens the door to the killing of the crippled, the aged, and those who are a burden on the community and the public purse.[17]   He therefore appeals to a conceptual distinction between "beneficient euthanasia" and the Nazi programme. This involves a further distinction between "killing out of kindness" and the "kindest way of killing". Quite obviously, if one is to kill it is better to do it kindly. But this does not suggest that killing kindly is just. One might embark on an evil programme of killing and still do it kindly. Killing out of kindness is clearly a different matter. There is no way in which the Nazi programme could be seen as one of killing out of kindness, argues Kohl, since it was a paradigm of cruelty, although some defenders of the Nazi ideology may have practised the art of killing kindly. But Kohl is concerned that many opponents of beneficient euthanasia believe that ultimately all advocacy of euthanasia rests on principles of utility. For this reason he goes to great lengths to distinguish between killing out of kindness and killing to eliminate various forms of human misery. In this respect there can be no objections to Kohl's rejection of "fiscal utilitarianism" and appeals to euthanasia in order to reduce the cost of medical care, when he warns that "the reduction of ethical principles to matters of economics generates fear and warrants the parade of Nazi horrors".[18]

We shall return to the arguments concerning fiscal utilitarianism, but this is the point at which questions have to be raised regarding Kohl's distinction between "beneficient euthanasia" and the "less desirable and immoral" fiscal euthanasia. For Kohl does not show how we can prevent the fiscal euthanasia programme from passing under the guise of "killing out of kindness". Now one form of the slope argument asserts that killing out of mercy leads to an increase in other forms of killing, not because of a logical connection, but because of the possibility of straightforward empirical confusion − naive or deliberate − between one form of killing and another.

For those who want to kill for one reason may be able to rationalize it with appeals to another reason. Those who want to kill a social nuisance will say that it was for his own good. A criminal sentenced to life imprisonment may come to believe he would be better off dead. Could we be sure that his death in such a case is not because of a desire on the part of others to be rid of a nuisance? No doubt there is less moral revulsion towards the notion of mercy killing. But how do we disentangle motives? As Lady Summerskill pointed out when the House of Lords rejected the Voluntary Euthanasia Bill of 1969, "undoubtedly there will be somebody to remind the invalid of his newly acquired powers over his own disposal".

The point of the slope argument is to demonstrate the practical difficulties in maintaining conceptual distinctions; to remind the erring Platonist that concepts are elastic and subject to change and manipulation. By drawing attention to a distinction between "fiscal" and "beneficial" killing, Kohl has not refuted the slope argument; he has only pointed to a distinction which his opponents would acknowledge. His failure concerns the crucial task of showing how the distinction can be maintained in practice.

So far, objections to Kohl's position have been based on empirical problems. However, objections based on logical grounds can also be raised. In a reply to Kohl, Arthur Dyck points out that "in dealing with the wedge argument, Kohl has not yet confronted it in its most powerful form. A wedge argument does not have to predict that certain practices will follow from another. A wedge argument is concerned with the form or logic of moral justifications".[19]

To appreciate Dyck's objections we have to examine Kohl's paradigm case of beneficient euthanasia: a child severely handicapped, not suffering pain, but nevertheless in a serious and irremediable physical condition which arouses in others a wish to help. In this case, argues Kohl, induced death would probably be considered an act of kindness by most people. But the logical objection which the slope argument raises is that even in this case it is difficult to draw the boundary of the category of beneficient euthanasia. In the foregoing example it would appear that Kohl's criterion is that of preserving the dignity of human beings. In Kohl's case this applies to a child born without limbs, sight, hearing, or a functioning cerebral cortex. Although not in pain and not dying, it would nevertheless meet Kohl's criterion. But some have argued that mongoloids, however happy or educable, are also lacking in dignity, so that they too, it would seem, could meet Kohl's criterion. What the slope argument is saying is that there is no clear and unambiguous way of restricting the category once the category for authorized killing has been accepted. Unlike arguments for killing in self defence, the arguments for beneficient euthanasia can extend to an indefinite number of cases, and the reasons for restricting them are not likely to achieve universal assent.

It may be significant to distinguish between a Nazi's motive for killing and beneficient euthanasia. But it is not merely a question of whether one society is kinder than another; there are problems as to what *counts* as an act of kindness, or what constitutes a meaningful and dignified life. Here we can have widespread differences which may be influenced by philosophical or theological assumptions concerning the concept of a meaningful life. Kohl's principle of killing out of kindness and concept of human dignity do not avoid the slope. Their logical implications are extensive — even if they may not reach all the way to genocide. For example, if we link the minimizing of suffering with killing, then we face the possibility that killing may be a preferable way of minimizing suffering than, say, the provision of companionship or long term care. But the very point of the slope argument is to prevent the ruling out of *other* means of alleviating suffering. It is invoked in this context because it is believed that once killing becomes an acceptable way of alleviating suffering, this could lead to a retraction of other means. Even if the logical implications of tying killing to the alleviation of suffering went no further than Kohl intends, this could unjustifiably pre—empt the objection that a request for euthanasia is a reaction to bad nursing.

Now Kohl does not wish to minimize suffering by a resort to killing. He only advocates killing out of kindness. But the logical implications of killing out of kindness are much wider than he envisages, and the link between killing and kindness may close off other options. For Kohl, beneficient euthanasia is a last resort. But if there were a general policy of beneficient euthanasia, with Government approval, actual practice might be very different. Last resorts, when legalized, have a habit of becoming first options, not necessarily because killing is always contagious, but because the concept of life not worth living is open to numerous interpretations. And this is the implicit value of Alexander's parade of Nazi atrocities. Even in a society where Nazism is not universally accepted, the concept of a life not worth living is unbounded.

IV Voluntary Euthanasia

The concept of voluntary euthanasia is often put forward as a counter to the slope argument. Thus Michael Tooley [20] combats it in his advocacy of euthanasia for a person who has a "rational desire that his life be terminated". For Tooley the philosophical problem is not so much a question of justifying voluntary euthanasia as one of examining the reasons why "many people view voluntary euthanasia as morally objectionable", and the belief that we have an obligation to go on living because of religious beliefs in a Creator is summarily dismissed by denying the existence of a Creator.[20] But the argument that attracts his attention is the slope argument, which he locates in the following remarks by G.K. Chesterton:

> Some are proposing what is called euthanasia; at
> present only a proposal for killing those who are
> a nuisance to themselves; but soon to be applied
> to those who are a nuisance to other people.[21]

In reply, Tooley argues that:

> If someone were to advocate sexual activity, and
> a critic were to object that while only voluntary
> sexual activity is being advocated at present, the
> proposal will soon be extended to cover compulsory
> sexual activity, i.e. rape, the critic would hardly

be taken seriously. If the analogy is a fair one, the objection is certainly preposterous.[20]

But is it a fair analogy? Apart from the quite valid objection that rape is not a form of sexual activity but rather a form of aggression, there is the rejoinder that in a climate of scepticism, and even hostility towards rape victims, the slope argument might have considerable significance in deliberations over the re—drafting of criteria for illegal sexual intercourse. Nevertheless, for Tooley, it is somewhat puzzling as to why intelligent people accept the slope argument. His explanation is that they fail to distinguish between a) arguments for the sanctity of life and b) arguments for the right to life; and that it does not follow that the rejection of a) entails the rejection of b). The right to life argument insists that every human being has a right to life, but it does not mean that it is always wrong to kill a human being. Moreover, the rejection of the sanctity of life principle, argues Tooley, can be done without any commitment to the view that decisions to terminate life should be based on concepts of social utility. One can argue for the right to life but acknowledge other rights, such as the right to terminate one's life, or the right to assisted suicide. Thus, for Tooley, the slope can be avoided once we recognize that we can maintain a commitment to voluntary euthanasia whilst supporting the right to life and opposing compulsory euthanasia.

This distinction is very important, but the slope argument returns when arguments based on social utility are employed by a supporter of voluntary euthanasia and the right to life. For example, suppose one says something along the following lines: "Although I respect the right to life, in my case when I am no longer socially useful I shall want to be put to death". In such a case the notion of social utility is part of the individual's criteria for the exercise of the right to assisted suicide. As such it can always be employed as a form of moral pressure which can be extended to the point where a person who wants to live may be placed in a position where he or she is obliged to justify the continuation of the life in question.

According to EXIT (formerly the Voluntary Euthanasia Society) voluntary euthanasia should be the "lawful right of the individual, in carefully defined circumstances and with the utmost safeguards, if, and *only* if, that is his expressed wish".[22] This proposal is based on an appeal to compassionate circumstances, and the society maintains that it does not entail "getting rid of the old, the infirm and the unwanted ..." nor "the 'putting down' of deformed children and mental defectives".[22] The significance of the slope argument here, however, is that it draws attention to the often insoluble problems concerning the determination of those "carefully defined circumstances" according to which an individual expresses a wish to die. It may be discourteous to suggest that among EXIT's supporters there could be those who favour a looser interpretation of voluntary euthanasia than that which is proposed, but this can never be ruled out. Similarly, it may be imprudent to draw any sinister conclusion from the fact that EXIT include a "form of wording of bequest" (with instructions for bequeathing a sum of money to the society) as an appendix to their pamphlet on voluntary euthanasia, but there are no logically compelling reasons why we cannot attribute an ulterior motive.

Even if we could be absolutely sure that the choice of accelerated death were purely voluntary, that is, not a response to expensive or inadequate care, pressure from relatives and so on, we still find problems over the rightness of assigning this degree of freedom to individuals. We do not regard restrictions on the ability to kill as a restriction of the individual's rights; why make exceptions when the life in question is the agent's? Sir Immanuel Jakobovits speaks for those who argue that euthanasia, however freely decided and however much it may be in the interests of the individual for the relief of suffering, should be resisted insofar as it may be connected with a general cheapening of human life. He resists euthanasia as a cause and a condition of a weakening of respect for life.

> Once we compromise the infinite worth of every human
> life and make any human life finite in value, turning
> it from being absolute into becoming relative — either
> relative to his state of health or relative to his
> usefulness to society — this will automatically bring
> about a situation in which some human beings will be
> worth more and others worth less, ultimately leading
> to the Nazi doctrine whereby human beings were graded
> and shoved into gas ovens by the millions because they
> were inferior in value.[23]

The right to die is a dangerous slogan. Too often it provides cover for those who seek the right to kill.

V Cost—Benefit Arguments

Notwithstanding his distinction between the right to assisted suicide and arguments for euthanasia based on concepts of social utility, Tooley ultimately supports the latter. His paradigm case is of a "person who wants to go on living, but the cost of keeping him alive is so great that letting him die is the lesser of evils".[24] This case, although individually distressing, argues Tooley, "is not morally objectionable".[25] Because "the first and least controversial point is that there are limited resources available for saving lives, and if the cost of keeping a person alive is too great, one will be able to save more lives either by diverting the resources to other patients or to medical research".[25] This argument must be rejected whenever it appears. First, it presupposes that a fixed sum is available and no wider influx of resources is possible. Second, it lends support to the socially divisive doctrine that wealthy people should live longer. Against the first presupposition it must be observed that from its inception medicine has been short of funds. If it were to have cut its cloth according to its size there would have never been enough finance for one hospital. The second presupposition raises the wider issue of social justice in medicine. It will not be dealt with here, although it must be stressed that inequality of access to medical care, on whatever grounds and however widely it is practised, is alien to the spirit of Hippocratic medicine.

A rather interesting convert to cost—benefit arguments for the termination of treatment was the former Archbishop of Canterbury, Dr Donald Coggan. Quoting the 1975—6 Health Budget of £4564 million, his remarks are a revealing insight into the mechanics of the cost—benefit argument. "The resources of the national exchequer are not limitless", argued Coggan, "and the prolongation of the life of one aged patient may in fact entail the deprivation of aid to others and even the shortening of their lives".[26] After acknowledging the possible abuses of legalized euthanasia, Coggan was nevertheless concerned with the escalating cost of health care: "But the awareness of these appalling abuses must not blind us to the realities of a situation the severity of which will not diminish but rather increase as the percentage of old people rises and, quite possibly, the extent of Government financial aid reaches a figure beyond which it cannot go".[27]   Having recognized a very serious problem Coggan then reveals how easy it is to slide down the slope, from upholding the physician's primary obligation to the care of the individual patient, to arguments based on social utility, and ultimately to the surrender of criteria for medical care to the economic dogma of the moment:

> The doctor has a responsibility — an accountability —
> to the patient and the patient's family under his
> immediate care. But he also has a further
> responsibility — to the Government or, to put it
> more personally but none the less accurately, to
> his fellow taxpayers who provide the resources to
> keep the National Health Service going.[26]

Note how ethical decisions are swiftly reduced to economic considerations, but the objection to Coggan's remarks is not that economic considerations are irrelevant to ethical decisions, but rather that they do not provide the limits to moral discourse. To base ethical principles on fiscal criteria is to ignore the transcendental and absolute nature of ethical imperatives, which, in an important sense, override reality. In the *Tractatus Logico—Philosophicus* Wittgenstein draws attention to the way in which ethical discourse is unconstrained by reality.

> If there is any value that does have value, it
> must lie outside the whole sphere of what happens
> and is the case. For all that happens and is the
> case is accidental.
> What makes it non—accidental cannot lie *within* the
> world, since if it did it would itself be accidental.
> It must lie outside the world.[28]

Later, having distinguished between ethical reward and punishment and punishment and reward in the usual sense of these terms, he reiterates his view that the scope of values and consequently, ethical duty is contingently related to the real. Says Wittgenstein:

> *How* things are in the world is a matter of
> complete indifference for what is higher. God
> does not reveal himself in the world.[29]

Wittgenstein's placing of ethics outside the limits of language and reality is not bound up with an appeal to intuition or mysticism, but is to draw attention to a fact of everyday experience; that ethical imperatives are neither limited by the facts of life or economic dogma masquerading as necessity. Economic reality, however, is not the kind of limit that should be compared with other limits, like the inevitability of aging or the limits which ethical heroes have ignored when they have disregarded the consequences of acting on a moral imperative. It is a matter of political choice. The sum of £5000 million is not an ultimate reality: it can be removed by a reallocation of resources and an act of political will.

It must be acknowledged that an individual physician is powerless against the reality of limited resources. But the limitations of the real do not coincide with the limits of moral responsibility. It is a feature of reality that the living conditions of the aged in certain inner cities are intolerable, but this is not a reason for accepting it. It is not the task of moral philosophers to alleviate moral anxiety with references to economic reality; it is not their task to relieve moral anxiety at all. Nor is it irrational to feel personal failure when our actions are circumscribed by financial restraints. Otherwise we could not explain why a hospital starved of funds will become demoralized and the staff will suffer from a *personal* sense of moral failure. Moral responsibility is independent of economic limitations, and for this reason cost—benefit arguments have no place in moral discourse. There is, of course, a great need for the redistribution of moral responsibility. The politicians and public must be shown to bear responsibility for the allocation of health resources. To be sure, on a day—to—day basis, physicians have to make decisions with regard to actual resources and for that decision they are morally responsible. But this does not make the moral burden exclusively theirs. Philosophers who formulate principles to match available resources are unwitting spokesmen for irresponsible Government agencies.

The problem is not alleviated by the introduction of market solutions of the kind proposed by Tooley's [30] advocacy of a "reasonable income" which will enable us to decide freely how much should be spent in keeping us alive. Apart from insoluble problems regarding what shall count as a "free decision" in this sphere, Tooley's proposal simply ignores the facts that wealth is publicly created and that health care is beyond the reach of any individual. Indeed, few individuals could afford the sum required to train one doctor.

Cost benefit arguments often appear as significantly important side issues which, nevertheless, hasten the euthanasia programme down the slippery slope. However, one area where an accelerated death may not lead down the slope is the right to refuse treatment. But it is only when such a decision is unhampered by cost benefit considerations that the slope can be avoided. The right to refuse treatment, even to the point of death, is not synonymous with assisted suicide, active or passive. In the first place, a refusal of treatment may not be accompanied with a desire to die. A refusal to accept intervention on religious grounds, even when fatal, need not be accompanied by suicidal intentions. The patient may be willing to undergo alternatives, may wish to recover, but not by the methods that have been proposed by the attending physician. An agent contemplating suicide, on the other hand, will not consider *any* form of treatment. From the standpoint of the physician the refusal of treatment, even when death is inevitable, does not involve complicity in an act of passive euthanasia, since it may be done against the offered advice and intentions. To be prevented from saving the life of a patient who refuses treatment may be unfortunate, even tragic, but with the possible exception of nagging doubts that one could have been more persuasive, it falls into the same category as not having been consulted in time, and so on. Euthanasia, even when passive, on the other hand, involves a decision on the part of the physician for which he or she is morally responsible.

It has been argued here that the slope argument cannot be invoked against the right to refuse treatment. In fact, the slope argument could be invoked against proposals to limit this right lest they supply a precedent for unlimited forms of medical intervention. The right to refuse treatment in the face of unlimited intervention and, possibly, the extension of life beyond the point where death with dignity is possible, needs to be distinguished from the refusal of treatment on cost benefit considerations. It is one thing to decide whether to continue resuscitation when life has ceased to have any meaning, but quite another to base one's decision on the cost of the assistance required. Nevertheless, it is surprising how often these two questions are conflated.

In a paper on the problem of organ transplants, Harold De Wolf refers to "extreme measures used by physicians in hospitals to extend the half—life of people reduced to a level of existence worse than death".[31] But then he goes on to conflate the argument for a dignified death with cost benefit considerations: "These measures", he says, "may often be paid for by survivors who have not ordered them and cannot afford them. The dying man may have provided carefully to arrange for a decent income to support his widow when he was gone. Yet now the base of that income and of the means of support for her meaningful life is used up in the heaping of indignation and pain upon him only for the unwanted prolongation of his dying".[31] This problem is a genuine one, but, nevertheless, a side issue in the argument whether or not one *ought* to extend human life. Changes in the social and economic structure could very well remove the financial burden from the individual's shoulders. There are grounds for arguing that a health system which places decisions of this kind on the dying is iniquitous. The redirection of a fraction of defence expenditure towards health care, or a change of attitudes towards dependents in favour of a more collectivist standpoint, would eliminate cost benefit considerations — although the tremendous moral problem of whether to preserve life at all costs would still be with us.

VI Voluntary Termination of Pregnancy and the Slippery Slope

One of the central questions in the abortion argument is "When does human life begin?" Nevertheless, there is something fundamentally mistaken about this question, insofar as it presupposes that there is a specific moment, or time t, when we can indicate the presence of human life or personhood. Yet attempts have been made to locate the beginning of human life at various intra—uterine stages. It is with reference to the difficulty of defining such a precise and morally significant moment in the gestation process that the slope argument enters into ethical discussions on the termination of pregnancy.

According to Joel Rudinow, the "extreme conservative" position regarding abortion is one which locates human life at conception. As such, the argument employed by conceptionists is a version of the slope argument, which Rudinow presents as follows:

> Birth is a morally insignificant event in the
> history of the born individual. As far as personhood
> and entitlement to treatment from the moral point of
> view are concerned, birth, which for the body is a
> mere change of environment, is no more significant
> than the first birthday. But once birth has been
> demythologized, as well it should be, we are on
> the slippery slope. For no particular point between
> birth and conception is a point at which the
> person/non—person distinction can be non—arbitrarily
> located, because the differences in development
> between any two successive intra—uterine points
> are so unimpressive. Consequently, we are forced
> to locate the beginning of human life at the point
> of conception.[32]

According to Rudinow, the problem with the conservative position is that it blurs the distinction between zygotes and persons. This is undoubtedly correct. But Rudinow's mistake lies in thinking that it is the conservative's employment of the slope argument that is responsible for this confusion.[33] On the contrary, the slope argument is not meant to establish that human life begins at conception: it is simply directed against a position which makes claims on behalf of any significant moment in the gestation process. If the conceptionist employs the slope argument to support conception as the point at which human life begins, he or she will be drawing from the slope argument something which is not in it. In fact, we find an example of this in Norman St John Stevas' argument that since there is no qualitative difference between the embryo at the moment of conception and the moment of quickening, the embryo must be considered a human being.[34] The reply to this argument is that a foetus is no more a person than an acorn is an oak tree, and that the killing of an acorn is not the same as killing a tree. However, this reply only corrects a mistaken application of the slope argument, which is not a defence of conceptionism but rather a means of drawing attention to both empirical and logical difficulties in the maintenance of conceptual boundaries around definitions of life and non—life. Nevertheless, unless it is settled decisively what shall count as a human being it is impossible to predict that liberal attitudes towards abortion will lead down the slippery slope. If, as the slope argument maintains, there is no significant event which marks the beginning of life, then there is no Rubicon to be crossed during embryological development upon which we can concentrate and say "Before this moment we have an object as trivial as a nail pairing; after this we have an individual human being to which we must reserve the full sanctity of human life".[35] Unless we have an idea of what it is that we are disposing of, then we cannot say that we are on a slippery slope. The slope argument only applies to human beings or cases where the foetus is seen as a human being. In this respect prohibitions of abortion may be perfectly compatible with doctrines which have no regard for the "right to life". The Nazis outlawed abortion and even made it a capital offence. "It was", as Potts says, "the philosophy that produced concentration camps that also carried out the last European execution for abortion".[35]

VII Conclusion

The significance of the slope argument is not so much in its appeal to the *status quo*, although it does serve as a warning against the difficulty of maintaining newly proposed boundaries. It is a recognition of what is correct in Hegel's insight into the "cunning of reason". In the absence of absolute knowledge and consequently absolute control over the consequences of our actions and decisions, we cannot afford to ignore the possible misuses to which our proposals for conceptual reform can be put. We can see the force of this kind of appeal in Veatch's objections to the killing of the dying, where an analogy between the universal prescription against killing the dying and the red light traffic rule is drawn.[36] It may not be necessary to stop at every red light. It may even be advisable to break the red light rule, say in an emergency when no car or pedestrian is present. In order to allow for these circumstances it might be preferable to adopt a weakened version of the red light rule: "stop at every red light unless the road is clear and you have very sound reasons for ignoring it". The problem with this new rule is that it would lend itself to numerous interpretations, and would multiply mistakes and misunderstandings, and create numerous borderline instances. It is therefore preferable to follow the existing rule, however inefficient. This situation is similar to proposals for accelerated death. The force of the slope argument is to reveal the problems entailed in proposals to redraft the rules in order to accommodate what are currently regarded as ethically justifiable breaches of it. Under the present laws a wise judge will show leniency towards a driver who breaks the red light rule in order to prevent a catastrophe. So would a wise judge be equally lenient when dealing with a case of accelerated death when there was overwhelming evidence that the killing was merciful. This requires no change in the law and would maintain a situation in which the onus is very much on those who wish to accelerate death to provide the justification for their actions.

The citing of exceptional cases where euthanasia might be morally defensible is no argument for a change in the law or for a re-drafting of our attitudes towards life. In such cases the law is best maintained, but with leniency and understanding.

Notes

[1]   Alexander, Leo, 1949, 'Medical Science Under Dictatorship',in *New England Journal of Medicine*, July 14, p.40.
[2]   Ibid. p.44.
[3]   Ibid. p.34.
[4]   Ibid. p.45.

[5]   Kohl, Marvin, 1974, *The Morality of Killing*, Peter Owen, London, pp.15−16.

[6]   Ibid. p.17

[7]   Ibid. p.19.

[8]   Ibid. pp.19−20.

[9]   Ibid. p.20.

[10]   Alexander, p.41.

[11]   Singer, p., 1979, 'Unsanctifying Human Life', in *Ethical Issues Relating to Life and Death*, ed. J. Ladd, Oxford University Press, New York.

[12]   Ibid. p.59.

[13]   Ibid. pp.59−60.

[14]   Kohl, p.98.

[15]   Mansson, Helge Hilding, 1972, 'Justifying the Final Solution',in *Omega*, vol. 3, pp.79−86.

[16]   Foot, Philippa, 1979 'Euthanasia', in *Ethical Issues Relating to Life and Death*, ed. J. Ladd, Oxford University Press, New York, pp.14−40.

[17]   Kohl, pp.96−97.

[18]   Ibid. p.99.

[19]   Dyck, A., 1975, 'Beneficient Euthanasia and Benemortasia: Alternative views of Mercy', in *Beneficient Euthanasia*, ed. M. Kohl, Prometheus Books, Buffalo, N.Y., p.120.

[20]   Tooley, Michael. 1979, 'Decisions to Terminate Life and the Concept of a Person', in *Ethical Issues Relating to Life and Death*, ed. J.Ladd, Oxford University Press, New York, pp.68−69.

[21]   Chesterton, G.K., 1937, 'Euthanasia and Murder', in *American Law Review*, vol.8, p.486.

[22]   *The Last Right: The Need for Voluntary Euthanasia*, revised edition, EXIT: London, 1980 pp.3−4.

[23]   Jacobovits, Sir Immanuel, 1974, 'Discussion on Death and Euthanasia', CIOMS 8th Round Table Conference, Geneva: WHO, p.142.

[24]   Tooley, p.71.

[25]   Ibid. p.73.

[26]   Coggan, Dr Donald, 1977, 'On Dying and Dying Well', The Edwin Stevens Lecture, Extracts, in *The Journal of Medical Ethics*, vol.3, pp.58−9.

[27]   Ibid. p.54.

[28]   Wittgenstein, L., 1961, *Tractatus Logico−Philosophicus*, Routledge & Kegan Paul, London, 6.41.

[29]   Ibid. 6.432.

[30]   Tooley, p.74.

[31]   De Wolf, L. Harold, 1973, 'Organ Transplants as Related to Fully Human Living and Dying', in *Ethical Issues in Biology and Medicine*, ed. Preston Williams, Schenkman, Cambridge Mass., p.38.

[32]   Rudinow,, Joel, 1974, 'On the Slippery Slope', in *Analysis*, vol.34.(5), pp.173−174.

[33]   Lindsay, Anne, 1974, 'On the Slippery Slope Again', in *Analysis*, vol.35.(2), p.32.

[34]   St John Stevas, Norman, 1963, *The Right to Life*, Hodder and Stoughton, London, p.32.

[35]   Potts, Malcolm, 1969, 'The Problem of Abortion', in *Biology and Ethics*, ed., F.J. Ebling, Institute of Biology, London, pp.74−75.

[36]   Veatch, Robert M., 1978, *Death, Dying and the Biological Revolution*, Yale University Press, New Haven, p.97.

8 Genetic improvement

RUTH CHADWICK

Suggestions for the improvement of the human gene pool, eugenics, are nothing new. Plato, in Book V of the *Republic*, put forward proposals for controlling breeding to ensure that the bravest men should have more chances to mate, in the hope that the stock would be improved. Earlier, in the late sixth century B.C., the Greek poet Theognis had lamented eugenic decline. What *is* new is the range of methods now available. Breeding programmes are now considered to be the 'old' eugenics.[1] The new eugenics might make use of any of the following: foetal screening, with the possibility of following it up with abortion; sterilisation of known carriers of genetic deseases; gene therapy arising out of recombinant DNA research, artificial insemination by donor (A.I.D.), cloning, and in vitro fertilisation (I.V.F.).

The new methods have provoked fresh arguments about the ethics of their use, but problems about the whole enterprise of eugenics still need further discussion. 'Master race' theories and Nazi policies have to a certain extent given eugenics a bad image, with the result that the very fact that A.I.D. might be put to eugenic use has been seen as one of its disturbing features.[2]

In this paper I propose to consider three aspects of the eugenics debate: first, the arguments against any form of eugenics; secondly, the distinction between different kinds of eugenics; and lastly, the means now in vogue.

I   Principle of Non—Interference in the Gene Pool

Those who object to eugenic policies and who hold a principle of non—interference in the gene pool might do so because they think it preferable to concentrate on improving the environment. I shall call one who takes this view an environmentalist. He might think in terms of changing the environment as being a policy carried out *for* people, while seeing genetic intervention as altering people to fit in with how we want society to be.[3]

Improving the environment would include raising standards of care for those suffering from genetic handicaps, rather than trying to bring it about that such individuals did not exist. But it would be wrong to think in terms of a simple choice between altering the environment and interfering in the gene pool. For changes in the environment can bring about changes at the genetic level. With regard to the species, environment influences which genes are selected for, and survive in, the gene pool. This explains, for example, the prevalence of the sickle—cell gene in areas where there is a high incidence of malaria. Those who are homozygous for the gene, inheriting it from both parents, suffer from sickle—cell anaemia, which is often fatal in early life, but those who inherit the gene from only one parent do not suffer at all, except that their red blood corpuscles become sickle—shaped. The advantage of being heterozygous for the gene is that it carries with it immunity to malaria, and this explains the frequency of its occurrence in malarial areas.

With regard to the individual, the environment is thought to play a part in activating genes, or 'switching them on'. Little is known about how genes are switched on and off, but it is clear that different cells in the body, having exactly the same genetic complement, perform different functions according to whether they are, e.g., in the liver or the kidney. So, if one considers the individual as a whole, it seems not unlikely that different genes in the genome may be activated according to the environment in which the individual finds himself. Thus Thoday writes, in pointing out the importance of both genes and environment:

> "No characteristic is largely acquired. Every characteristic ... is entirely acquired. Every character of an individual is acquired during the development of that individual. Likewise every character is genetic, for to acquire a character during development in any particular environment the individual must have the necessary genetic endowment."[4]

It is impossible, therefore, to draw a sharp distinction between changing the environment and intervening at the genetic level. In fact, what has given the eugenics movement its impetus is precisely the claim that the changes that have been made in the environment, particularly in the twentieth century, have led to deterioration in the gene pool. We *have* been altering the genetic quality of the species, and have been doing it in the wrong direction. In Theognis' day, the lament over the decline in the breeding stock laid the blame at the door of those who wanted to marry for money rather than for good heredity. Today the problem is said to lie in the advances in medical care and the rise in living standards, which together have made it possible for those who previously would not have survived long enough to reproduce to do just that.[5] As a result the species may be growing progressively weaker from a genetic point of view.

So the environmentalist cannot draw a firm distinction between changing the environment and interfering in the gene pool, but he may still argue that it is preferable to pursue policies aimed at improving the environment, which may have indirect effects at the genetic level, than deliberately to pursue a policy of eugenics. For eugenics raises the following problems:

(a)   Who will benefit?

Why should we care specifically about the human gene pool? It is difficult to grasp exactly what it is that we are required to be concerned about. At least if we concentrate on improving the environment, we may be able to see results for people here and now, and if these are good then the possibility of indirect effects on the gene pool seems to recede into the background.

Eugenists worry about the human gene pool because of the consequences for the future of the species. But should it be of concern to any given generation that the species may not survive? Perhaps it is asking too much to hope that people will think in terms of the good of the whole species. A 'eugenic principle' of the form 'Always act in such a way as to promote the survival of the species' would be unlikely to gain much support. J.L. Mackie has used the term 'ethics of fantasy' for moral views which demand such universal benevolence.[6] But even if we accept that the species is composed of individuals, there is still a difficulty. For it is future, as yet unactualised, individuals who will benefit, or not, from the eugenic policies we carry out now.

The first thing to remember is that this problem is not specific to eugenics. Many issues relating to our treatment of the environment, e.g. pollution and the conservation of natural resources, share this feature.

It is beyond the scope of this paper to discuss the arguments for and against the view that we have obligations to future generations. The interesting thing is: if we do, does this include an obligation to be concerned about their genes? An environmentalist might want to say that the obligation ends at caring about the environment. This might be because of the nature of the benefit to future generations or the nature of the sacrifice to the present one.

(b)   The nature of the benefit.

When we talk about conserving natural resources and decreasing pollution, we are thinking in terms of what we ought to do for people who will exist in the future. When we speak of eugenics, however, we are tackling the question of how we can determine what those future people will be like. To say 'whoever will exist in the future, we should leave them a planet in as good a condition as possible', is very different from 'we should try to ensure that the people who will exist should have characteristics x, y and z'.

The eugenist could argue however that these two are not as different as they appear. Why do we want to leave the planet in as good a condition as possible for future generations? One reason might be that we do not want them to be unable to survive because of our selfishness. Another might be that we do not want them to have a very low quality of life. If what we are really concerned about is the survival capacity and/or the quality of life of future generations, and we find that genetic capacity has a bearing on these, then that gives us a reason to carry out eugenics. If poor genetic quality adversely affects them, while good genetic quality enhances them, then anyone who supports their promotion has reasons to support genetic intervention.

It might be doubted that genetic quality has a connection with quality of life, at least. For every example of an individual whose life has apparently been blighted by his shortcomings from a genetic point of view, one could find another whose fortitude has enabled him to overcome his handicaps and live a happy and contented life.

Arguments over this question are inconclusive, as we can see from disagreements over whether handicapped babies should be allowed to die. But let us suppose that it is granted that there is a connection between genetic capacity and quality of life. It might still be held that the sacrifice involved in eugenics is too great.

(c)   The nature of the sacrifice.

One might think that the extent to which individuals in one generation will be prepared to make sacrifices for superior future genetic quality will depend, not only on what they actually have to do, but also on the extent to which they will feel able to identify with the future beneficiaries. In the case of the conservation of natural resources we are required to imagine what it would be like for people similar to ourselves to be without them. Eugenics, however, might produce people with whom we could not easily identify. Books with titles such as "Our Future Inheritance: Choice or Chance?"[7] attempt to bridge the gap by encouraging us to think of the future of the species as something that belongs to us.

It is, at least, possible that one could identify with someone suffering from genetic disease, and that this would move us to some kind of eugenic enterprise. Further, it seems clear that one can expect people to be concerned about the immediately following generation, their own children. For one thing, they are going to have to live with them, and they may be dependent upon them when old. If it is claimed that this concern does not extend to worrying about their genetic quality, one can point not only to the effort that may be required in looking after a handicapped child but also, these days, to the threat of a 'wrongful life' suit.[8]

So there are good reasons for members of the present generation to worry about the genetic quality of their own children. The eugenist's best bet, then, is to frame his eugenic principle not in terms of the species but as, e.g., 'Other things being equal, it is better, morally speaking, to produce a child with a higher, rather than a lower, genetic potential'. Such a principle might gain wide support.

It is important to note that the principle does not necessarily urge producing a child with the highest possible genetic potential. It is to be admitted that eugenics is unlikely to be successful if there is too great a gap between one generation and the next. Parents may not want to feel completely inferior to their children.[9]

Having established that there are prudential reasons for being concerned about the genetic quality of one's children, quite apart from any moral considerations, there is the further question of whether the particular actions, or restrictions, involved in eugenics would demand too great a sacrifice. This would depend very much on the particular programmes carried out. No general argument could be made out to support the conclusion that eugenics demands too much, now that the range of methods is so extensive. It may have been the case in Plato's day, but not now.

(d)   Risk.

There has always been a general problem concerning eugenics, that there is a risk that we may not like the results of any policy that is carried out. We cannot be sure that we shall achieve the results we want. As we can see from eugenic policies carried out in animal species, what is gained in one direction is often lost in another. "You cannot breed racehorses that are also good carthorses."[10] There is the danger then that in aiming to control the characteristics we want in the gene pool, we may inadvertently lose others we value.

Now, of course, particular aspects of the new technology all have their own risks too.

The eugenist might reply that, consistently with what has been said about the generation gap, we are not going to attempt to do anything that will make too radical a change between one generation and the next, even if that were possible. That should reduce the risk.

Secondly, eugenists may argue that genetic diseases, such as Huntington's chorea, cause such misery to their sufferers that it is worth taking even quite substantial risks in order to eliminate them. If we do nothing, we have the certainty of a fairly high percentage of people suffering from genetic defects [11], whereas if we adopt genetic policies we have at least the possibility of lessening the frequency with which they occur. But, accepting that there are risks, the more sensible approach may be to assess each of the means individually by some such technique as cost—benefit analysis, rather than write off the entire eugenic enterprise. It is conceivable that circumstances may arise in which a refusal to countenance eugenic policies may carry the greater risk, viz. extinction of the species.

II   Principle of Limited Interference

Some writers on the topic of eugenics feel that it may well be worthwhile to carry out eugenic policies in order to reduce the frequency of genetic diseases which cause suffering, but point out that not all of the eugenists' proposals fall into this category. Some go much further than this.

There are those, then, who want to draw the line at this point and make a distinction between 'negative' and 'positive' eugenics. Bertrand Russell and the Medawars, for example, have held that positive eugenics is not feasible, whereas negative is; both hold that positive eugenics is not politically feasible, the latter also that it is not genetically feasible.[12]

Catherine Roberts, too, states that

"the eugenists' proposals which come under the heading of 'negative eugenics' and which include reducing man—made radiation and discouraging genetically defective individuals from reproducing ... all of these have my sympathy"[13] but she denies that there are "any convincing arguments" by which the dangers of positive eugenics can be minimized.[14]

Before considering these claims it is necessary to be clear about what the difference between negative and positive eugenics is. They differ in their aims: that of negative eugenics is the removal of, or lessening in frequency of, genetic defects in the gene pool. Positive eugenics aims at improving the quality of the gene pool, so that the species will reach new heights. (This is intended metaphorically, but a eugenist might of course mean it literally.)

The distinction is not entirely clear, as it is not immediately obvious where removing defects ends and making improvements begins. Any lessening in frequency of deleterious genes will at the same time be an improvement in the quality of the gene pool. We need a clear understanding of what is to count as a genetic defect, and what is to count as an improvement, and this is extremely hard to come by.

(a) What is a defect?

One element in the claim that positive eugenics is not genetically feasible is that there are no clear criteria for genetic improvements, the implication being that genetic defects are understood — at least, we know what they are. We must examine whether this is the case.

It might be thought that one approach to this question would be to establish a norm, and then say that all who fell short of the norm were defective, while all those above it were especially well endowed genetically. Unfortunately, as Tranöy points out, it is difficult to know where to begin in trying to state the necessary and sufficient conditions of normality.[15] It may nevertheless be possible to establish a norm for any single characteristic, e.g. chromosome number. This could be done statistically.

Problems arise however over whether we are considering the species as a whole, or a particular population in its specific environment. Characteristics which may be 'abnormal' statistically over the entire species may be possessed by the majority in a given population, e.g. the sickle cell gene mentioned above; so the eugenist has to consider the extent of his ambitions. In *Brave New World* the World Controller has wider powers than one can envisage for any present day eugenist.

We might suggest, as a first condition that a characteristic must satisfy in order to count as a genetic defect:

(i) It must be, in principle at least, attributable to some feature of DNA, such as a single malfunctioning gene, or a chromosomal aberration. We understand e.g. whether a gene is malfunctioning by reference to the 'normal' performance of that gene in the remainder of the population.

However this, while it may be necessary, is not sufficient to establish a characteristic as a genetic defect. For if evolution works via random changes that spread through the population if they prove advantageous, their statistical abnormality when they first arise cannot reasonably justify calling them defects. So we also need:

(ii) A genetic defect must confer some kind of disadvantage. What kind of disadvantage?

(a) One possibility would be disadvantage in quality of life, but, as we saw earlier, it would be possible to disagree that genetic defects do lead to a disadvantage in quality of life for those who possess them. It has to be admitted that, whilst a connection between the two may exist, it is nevertheless a contingent one, and can therefore play no part in the definition.

(b) The second possibility is biological, or genetic, disadvantage. Genetic defects are disadvantageous to those who possess them, it might be claimed, because those individuals will do less well in evolutionary competition. They will be less successful in reproducing and passing on their genes to offspring.

Are (i) and (ii) b) both necessary and jointly sufficient for a genetic defect?

There are considerable difficulties in the way of accepting this. There is the problem that genetic disadvantage to the individual has been counteracted by advances in medical care, as we saw earlier, so that people with genetic defects in many cases do not die before they have the chance to reproduce.

It is useless to reply to this by saying that it is still very rare for those with, e.g., Down's syndrome to reproduce. That leaves the point of the objection untouched. For there are still genetic diseases which have a negligible effect on one's ability to reproduce.

Perhaps, however, we are looking for the disadvantage in the wrong place. To think of genetic disadvantage as accruing to the individuals only makes sense if one is thinking in terms of the selection operating at the level of the individual. But if selection operates at the level of the species, or even of the gene, then presumably genetic disadvantage would work at that level too.[16]

Criterion (b) might be reformulated so that the disadvantage accrues to the species. As more and more people pass on defects to their children the genetic load carried by the species would become intolerable, so that more and more people would be defective, until enormous resources would be required to counteract it.

Or it might be stated in terms of the gene. The gene involved in a genetic defect would be at a biological disadvantage because it would not do well in competition with other genes.

The problem with such reformulations is that it seems counterintuitive to speak of an individual as having a genetic defect which may not confer genetic disadvantage on him but which may be disadvantageous to the species as a whole or to the gene itself.

We should remember however that it seems counterintuitive because we see the eugenist's aim as to improve the chance of life and quality of life for individuals. If so it may not matter to the negative eugenist that he cannot state precisely where their genetic disadvantage lies, as long as he can produce a list of genetically produced conditions he wants to eliminate because they reduce the quality of life.

It may be the case, then, that in saying what counts as a genetic defect the eugenist will be making an evaluative judgment rather than a descriptive statement. But if he wants to call himself a eugenist he cannot afford to ignore questions of genetic advantage and disadvantage altogether, because a eugenist is by definition one who wants to improve the human gene pool, even if he wants to do this in order to improve the quality of life.

Why not speak in terms of genetic diseases rather than genetic defects? Motulsky points out that a condition such as cleft lip and palate is a birth defect in which genetic factors play a part, but we might not want to class it as a disease.[17] (And yet, presumably, we might think it desirable to eliminate it.) This implies that there are some criteria for what counts as a disease which are lacking in the case of some things we should want to call genetic defects.

But similar problems have arisen in attempts to define disease. P.D. Toon, discussing suggested definitions in terms of conditions of abnormality plus biological disadvantage, points out that doctors try to cure diseases because they are concerned about the welfare of individual patients, not evolutionary problems, just as eugenists want to improve genetic quality in order to improve quality of life.[18]

It has been argued, in the case of disease too, that the concept is at least partly normative. The fact that homosexuality for example, was once considered a disease is cited as evidence for the view that the concept of disease is applied to conditions that are disapproved of in society.[19]

It may not be as simple as this, but it seems to be the case that the notions of 'defect' and 'disease' are far from clear. What I have tried to show is that negative eugenics is by no means on totally solid ground when criticising positive eugenics for lack of clear criteria.

(b) Criteria for improvement.

We might begin by considering some of the candidates that have been proposed as possible genetic improvements: higher I.Q. and greater moral sympathies are two favourites.[20]

One objection to such proposals is that in order for something to count as a genetic improvement it must at least be attributable to genes, as genetic defects are. But those characteristics suggested by positive eugenists are not clearly genetically determined at all. The debate over whether I.Q. is hereditary or genetically determined is notorious. Further, even if it is genetic, it may be multifactorial, and therefore very difficult to improve by eugenic policies.

Secondly, it might be the case that we could introduce all sorts of changes that we think would give individuals higher quality of life, but in addition to this we have to keep in mind the question of genetic advantage and disadvantage, as the negative eugenist does.

To give an example, E.O. Wilson has considered the possibility of removing the incest taboo, which he claims is genetically determined, so that people would feel more relaxed about incest. It could be seen as just another sexual barrier being broken down. But Wilson says that however attractive we might find this idea, if we do, we have to consider the genetic price we should have to pay, which could be considerable.[21]

It is no good, we might say, for the positive eugenist simply to concentrate on the characteristics he likes the best, from whatever point of view, or on those he thinks will improve quality of life, while ignoring whether these changes would be eugenic or dysgenic.

But as we have seen negative eugenics is little better off on this score.

Anyway, it does not seem theoretically impossible for the positive eugenist to come up with a list of improvements that would satisfy these demands, if the negative eugenist can produce a list of defects. Perhaps it will have to wait for greater knowledge about which qualities are genetically determined, and to what extent.

(c) Strength of obligation.

The second argument against positive eugenics might be that if we have any obligation to intervene in the gene pool, it stops at the level of negative eugenics. We may have an obligation to decrease the amount of suffering in the world, and thus to try to cure genetic diseases, but there is no such obligation, or a far less important one, to make improvements for those not suffering from them. Even had we established what the improvements might be, why should we take the trouble?

Arguments for such views are familiar. There are those who support, for example, a distinction between negative and positive duties, saying that our duties to refrain from harm are stronger than our duties to save people from suffering, and certainly stronger than duties to improve the lot of those who are not in need.[22] Those who take this view might even refuse to support certain forms of negative eugenics, if they did not take the view that bringing a handicapped child into the world could count as harming him.[23]

Anyone who holds the eugenic principle thinks that, other things being equal, it is better, morally speaking, to bring into the world an individual with a higher, rather than a lower, genetic potential. This would apply whether negative or positive eugenics was under consideration. We said earlier that such a principle might well gain wide support. On what grounds?

Some forms of utilitarianism might provide arguments for such a view. Let us consider the sort of reasoning employed by Jonathan Glover in *Causing Death and Saving Lives*. He is arguing for a principle that we ought, other things being equal, to maximise the amount of worthwhile life, but he also hints at the possibility that it would be right to maximise the worthwhileness of life, and he says:

> "being glad to be alive is quite compatible with the recognition that, from an impartial point of view, it would have been even better if some more gifted or happier person had been conceived instead."[24]

So in the case of artificial insemination by donor, let us make the wild supposition that our technique for analysing sperm had become so sophisticated that we could tell that, given two alternative samples, there was nothing to choose between them except that the one carried genes known to have some correlation with being musical. Why not choose that for that reason, supposing that being musical were one of the items on our list of desirable characteristics? Would it not be better, morally speaking?

It is difficult to see how a case could be made out for its being morally wrong, at least given that the practices of A.I.D. were thought morally acceptable in the first place, unless special circumstances were assumed, such as a glut of musicians. But those who were unsympathetic to utilitarian views might hold that the choice was morally indifferent. Anyone who holds the eugenic principle, however, should take the view that the introduction of the musical gene would be better, morally speaking.

It might be argued, however, that negative policies should always take priority over positive policies. Negative utilitarians, for example, might try to uphold a minimum acceptable level of well−being, and say that we should concentrate our efforts on raising everybody to that standard before we attempt to make improvements for those above it.

But where could such a minimum acceptable level be placed? If it is put at that point below which life is not worth living, it is far from clear what severity of genetic defect would place an individual in that category. Moreover, having such a minimum acceptable level would not indicate an obligation to try to find a remedy for many genetic defects, since they do not make life such that it is not worth living, although those who suffer from them may prefer a life without them. But we do not adopt such a policy with diseases of the phenotype. We try to cure all diseases, whether they make life intolerable or not.

On the other hand, to place the level at that point where negative policies end and positive policies begin would be too high. It would be extremely optimistic to suppose that we should ever reach a stage when no defects were present. New mutations and inadequate methods of screening make this an unrealistic ideal. We cannot place a minimum acceptable level at an unreachable point.

Moral views which depend upon the notion of a minimum acceptable level, or negative utility, have severe difficulties.[25] Of course, one who supported the eugenic principle on utilitarian grounds might have to admit that by the principle of diminishing marginal utility it would be less likely that significant improvements in quality of life would be achieved by positive eugenics than by negative, but this is not an argument against attempting to make an improvement by a positive policy should this prove possible, nor is it an impediment in the way of negative policies.

III   The Means

Negative	*Positive*
discourage gentically weak from reproducing	encourage genetically well endowed to have children
compulsory sterilisation	compulsory breeding programmes
foetal screening and abortion of defective foetuses	artificial insemination
infanticide	I.V.F. + genetic engineering
gene therapy	cloning

The chart is intended as a general guide to methods of controlling genetic quality. It is not exhaustive, and I do not propose to discuss all of the methods individually. Four general points need to be made, however.

First, at the level of the chart consisting of 'discouragement' on the negative side, and 'encouragement' on the positive, it should be pointed out that these can include counselling, propaganda, and tax incentives. It seems convenient, for the purposes of the chart, to bracket them together.

Secondly, A.I.D., I.V.F. and cloning have been placed on the positive side, but they could also be put to negative use. Cloning for example could conceivably be employed in the case of a couple where one partner had a serious defect which he or she is reluctant to pass on. The couple could have a child by cloning the other partner.

Thirdly, one thing that is clear is that it is not the case that all negative policies are morally acceptable, while positive ones are not. Those, like Catherine Roberts, who support the former against the latter sometimes suggest that this is so. Having expressed her support for negative eugenics without even discussing the means, she writes of positive eugenics:

> "All of the serious problems regarding paternity and maternity (the successful engrafting of deep frozen ova into women is anticipated) and the overthrow of traditional social conventions are too obvious to discuss."[26]

This was written before test tube reproduction became a reality. The Medawars' claim that positive eugenics is not politically feasible could also be interpreted as an expression of the view that there are no acceptable means of carrying it out. If it is thought that it should be interpreted as a point about the social divisiveness of suggesting that certain individuals are superior to others, it is not clear why negative eugenics should escape this criticism. The same type of decision, about what is better and what is worse, what should be encouraged and what not, has to be made in negative eugenics. But it is important to question the validity of this criticism. Perhaps the decision should be understood as one concerned with what genes to encourage, not which individuals, so that eugenics need not be seen as showing lack of respect for certain individuals. The two must be kept distinct, but some of the means make it very difficult to do this. Whether the 'new' eugenics makes this easier is something that must be considered.

Finally, in what follows, although I shall speak of eugenics in general, I shall primarily be concerned to discuss whether there is a morally acceptable means of positive eugenics, and I shall not be discussing, for example, amniocentesis and abortion.

When we look at the more traditional methods first, the level of compulsion seems unlikely to be acceptable for either negative or positive ends, because it involves gross interferences with liberty and privacy. It seems clear that this could not be feasible either politically or morally.

When we turn to encouragement, however, what about financial incentives, also associated with the old eugenics?[27] Such measures involve discrimination in that they not only suggest that some adults are more suitable as prospective parents than others, but also offer them advantages. Incentives, then, seem inextricably linked with discriminatory practices with regard to existing adults.

This applies to counselling too, which on the face of it could surely be put to both negative and positive use. If it is used, as it is at present, to advise prospective parents on how to avoid defective children, it is difficult to see why counsellors could not advise on how to have the best child possible, genetically speaking, under the circumstances. The problem is that even counselling may be thought to involve discriminatory judgments about which adults should reproduce.

So let us turn to the new means and see if they avoid these objections. They include A.I.D., *in vitro* fertilisation, and cloning.

The first point to be made about A.I.D. and I.V.F. is that they were designed as remedies for infertility. An opponent of eugenics might nevertheless want to support their primary use. We must examine whether this view is a tenable one.

Let us start by giving some thought to A.I.D., since this is actually being used in an attempt to carry out positive eugenics at the present time. The "Repository for Germinal Choice", in Escondido, California, collects sperm from men considered to be outstanding in some way: Nobel prizewinners are a favourite source. Several women have been fertilized with sperm from this "Repository", in an attempt to produce 'superbabies'.

The "Repository" would apparently welcome deposits from Robert Redford, Paul Simon, and Paul McCartney. There is thus some attempt to preserve variety in its selection of donors of sperm, a must for any eugenic programme. Not all the donors are outstanding in the same field.[28]

What are we to say about this from a moral point of view? It is important to distinguish the following questions:

(1)   If these procedures are used at all, is it possible to avoid making eugenic decisions?

If we operate any kind of screening or selection of donors, then we cannot avoid eugenic decisions, and in fact there are moral arguments in favour of making them. If it is morally acceptable to use A.I.D. at all, to help couples of which the male partner is infertile, then it is difficult to see why it should not be used for some eugenic purposes in these cases. For example, it may seem obvious that sperm from individuals with serious hereditary defects should not be used, and in fact donors are checked for such characteristics. This is nothing other than negative eugenics.

If that is the case, it is not clear that there are any arguments for limiting these decisions to negative eugenic ones. If it is acceptable to bar the use of sperm that is of poor quality, why not use especially good sperm, the best one can get?

There seem to be two possible replies to this, given the presumption that A.I.D. is morally acceptable. One would be to argue that since we do not necessarily screen parents who are reproducing without any outside help, we should not screen sperm donors. But if the sperm is being provided by a practitioner who is in law offering a service, he might be liable in tort if he supplies sperm that produces a defect.

The second is to argue that A.I.D. should not be used in overt positive eugenic programmes, at least, because this would aggravate problems that are already raised by the practice when used as a remedy for infertility.

(2)   *Would* the use of A.I.D. for eugenic purposes aggravate problems already recognised as raised by A.I.D.?

First there is the effect on the child so produced. Much concern has been expressed over the fact that A.I.D. children feel cheated over not knowing who their biological father is. The legal status of the A.I.D. child is also uncertain. Then there is the relationship between husband and wife which may be put under strain. It is feared that the practice might contribute to the break—up, not only of individual families, but of the family as an institution, especially if single women start having A.I.D. children.[29]

No argument either for or against the family will be presented here, but it is worth noting that eugenics has traditionally been associated with the abolition of the family. It may be thought that the other problems can be overcome; e.g. by clarifying the law to solve the legal questions, or that they are thought to be acceptable risks to take in order to help childless couples.

But when we stop talking about the problem of childlessness, and turn to thinking of positive eugenics, a new dimension may be introduced. For according to the eugenic principle it is better, morally speaking, to have a child with a higher, rather than a lower, genetic potential, other things being equal. But this principle would appear to generate the conclusion that in the case of couples where the husband is not infertile, and where he has no genetic defect, it would still be better, morally speaking, to have a child with exceptional donor sperm than by him.

There may however be several reasons for thinking that other things are not equal here. In A.I.D., the husband is being denied the chance to pass on his genes. While argument would be needed to show that he has a right to reproduce, he may have the desire to do so and this is important. If the practice of A.I.D. for producing 'superbabies' became widespread, large numbers of women could have their eggs fertilised by a small number of men while the majority of men would be thwarted. It seems *prima facie* unlikely that this situation would produce a socially desirable result.

Secondly, in cases where A.I.D. is carried out it is common for the woman to have fantasies about the donor. This might be aggravated by the knowledge that the sperm came from an individual regarded as genotypically or phenotypically exceptional, with the possibility of added feelings of inadequacy on the part of the husband (where there is one). As A.I.D. is practised now, efforts are sometimes made to find a donor similar to the husband. Using it for positive eugenics might only accentuate the differences.

R. Snowden and G.D. Mitchell describe how great efforts have been made by couples to keep the details of the child's conception secret, both from the children themselves and from relatives and society at large. (It is not always clear whether this is to protect the child or the husband.) They consider this secrecy to be a bad thing because it produces an atmosphere in which the child is bound to sense something wrong, and in which relatives are being treated dishonestly, thinking they have, e.g. a niece, when in fact she is no blood relation.[30]

It is not clear whether those who have A.I.D. children would feel more inclined to be less secretive about the use of it to produce super—children, and what effect this would have. Would the knowledge that one was one of these children compensate for the problems of coming to terms with the unusual circumstances of one's conception? Of course, if the use of A.I.D. became widespread the circumstances would not be so unusual; but the more it is practised, the greater the threat may be to family life, and to male security.

So even one who supported the eugenic principle on utilitarian grounds would have to admit that A.I.D. as a positive eugenic tool could lead to the dissatisfaction of large numbers of people. It might be suggested that this could all be avoided by restricting the use of artificial insemination to cases where the undesirable effects are likely to be minimal. It could still be a eugenic tool, but without donor involvement. For example, sperm could be stored by young men who wished to postpone having children until later. Reducing the parental age has a eugenic effect: storing sperm taken from young men would reduce the risk of mutations that occur with increasing age. Thus it might be argued that it is better, morally speaking, for a woman to be fertilised with sperm stored by her partner before they were married, than with the sperm he produces now. A similar argument might be made in the case of those whose work entails exposure to radiation.

(3)   Would A.I.D. avoid the problems of the old eugenics?

We saw that the main problem concerning the old eugenics was that it involved discrimination. It is not clear that A.I.D. could avoid this charge. For example, if men considered to be 'desirable' were offered generous incentives to become donors, then there would appear to be very little difference from older methods. And we have seen that the more widespread the practice became, the greater the likelihood that fewer and fewer men would have the chance to reproduce.

(4)   Would it be effective as a positive eugenic tool?

Overt attempts to practice positive eugenics by A.I.D. are rare. On the whole it is advocated as a remedy for childless couples. What I have tried to point out is that, even if this is the case, it is difficult if not impossible, once it is to be considered acceptable, to avoid eugenic decisions. But if we did decide deliberately to practise A.I.D. for eugenic purposes on a wide scale, it has to be admitted that it promises low probability of reward. For there is no guarantee that the combination of donor sperm with the woman's egg will give us a desirable result, or one which is appreciably better than that achievable by normal sexual reproduction. The women who choose to be fertilised by Nobel prizewinners may be very disappointed by the results. The method currently being used by some for positive eugenics, then, seems unlikely to be effective.

So where do the alternative possibilities lie? According to Robert L. Sinsheimer, the greatest hope lies in genetic engineering arising out of recombinant DNA technology. From this comes the possibility of gene therapy for genetic diseases. New genes could perhaps be introduced into the bodies of those with genetic defects. He says

"The new eugenics would permit in principle the conversion of all the unfit to the highest genetic level".[31]

As regards positive eugenics, he suggests

"The horizons of the new eugenics are in principle boundless — for
we should have the potential to create new genes and new qualities as
yet undreamed".[32]

It might be the case that these techniques could be used in combination
with *in vitro* fertilisation. The suggestion might be that, when an egg is
fertilised *in vitro*, some genetic engineering might take place at this stage in
order to manipulate the kind of genes we should like the baby to have.

On the one hand this technique has advantages over A.I.D., in that both
partners in a given couple can contribute their genes. They might then be
able to opt for certain adjustments to their genetic mix, with or without
some centralised control over the range of options available to them. (One
problem for example concerns the management of the sex ratio, if an
overwhelming majority of parents desired boy children.)

The ethics of *in vitro* fertilisation is of course controversial. But the
question that concerns us is: if it is acceptable to help childless couples, is
it acceptable for eugenic ends?

Let us ask, as we did when considering A.I.D., whether it is possible to
avoid eugenic decisions if *in vitro* fertilisation is used. Suppose a
practitioner discovered that of three embryos fertilised *in vitro*, one was
carrying genes indicating that s/he would suffer from Huntington's chorea.
Anyone who thinks that it would be morally preferable to return one or
both of the other two embryos to the womb, rather than this one, is
implicitly supporting the eugenic principle. And if this decision is acceptable,
why not make the same sort of decision if that embryo, alone of the three,
was found to lack genes for, say, musical ability?

Some people's reactions to this example may be influenced by their
understanding of what is going to happen to the embryo that is not
returned to the womb. Some take the view that no embryo should be
produced which is not going to be returned to the womb; others think it
acceptable that embryos should be experimented on. This of course depends
on what view is taken of the status of the embryo.

But let us modify the example slightly, so that we are planning to give
all three of the embryos the chance of development, but we find that one
has this genetic make—up which we consider defective. Let us suppose that
fortunately, with new found knowledge of genetic engineering, we find that
we can modify the genes of this embryo to give it a higher, rather than a
lower, genetic potential. In other words, we take a negative eugenic
decision. So why not take a positive eugenic decision and add a few extra
capacities?

(5)   Would this aggravate problems already raised by the practice of
       I.V.F.?

It has already been pointed out that one central problem in this area is the
status of the embryo. But this is not a problem which arises only when we
are talking about the eugenic use of *in vitro* fertilisation. It is a problem

that arises in any case. Decisions have to be made about how many embryos to return for eventual birth, about what is to happen to the 'spare' ones. This eugenic aspect simply makes clear some of the criteria that might be used for those decisions.

(6)   Would these means avoid the problems of the old eugenics?

If it is thought that embryos should have the rights of persons, then these decisions might be thought objectionable on the grounds that they involve deciding against certain kinds of persons, i.e. that they are discriminatory. But if it is not thought that embryos have the same rights as persons, then these means might seem to avoid the problems of the old eugenics. Decisions are simply being made as to what kind of genes parents would like their children to have.

(7)   Would it be effective?

As Sinsheimer has pointed out, this method seems to offer the greatest potential for choosing the design of future human beings. The problem is how the choice should be manipulated. I have spoken in terms of parents choosing higher genetic potential for their children, but it has to be admitted that the accumulation of a large number of parental choices might lead to a bad result overall, from some points of view. But the prospect of control of choices by governmental or medical authorities takes us back to the problems of the old eugenics.

Cloning I have discussed elsewhere.[33] For the present I shall confine myself to the point that, even if it were to become practicable in human beings, it is unlikely to be a useful means to eugenic ends, as it preserves types already in existence rather than allowing for any improvements.

Conclusions

What I have tried to argue is, first, that eugenics, although it has a sinister image, may be interpreted as just another manifestation of wanting the best for children. The eugenic principle tells us that, other things being equal, it is better, morally speaking, to have a child with a higher, rather than a lower, genetic potential.

Secondly, I have argued that many people implicitly accept this up to a point, in that they are willing to countenance negative eugenics, but they want to draw the line at that point. This is an untenable position.

Thirdly, I have tried to show that moral problems arise in connection with the means of eugenics rather than the principle. The old methods are clearly discriminatory. Of the new means, I.V.F. seems to offer the greatest hope of avoiding these, and the most effective method too. (But the gravity of the moral problems arising in connection with I.V.F. and the new means may be argued to be greater than those of discrimination associated with the old.)

One point I want to stress is that if we are to use the new reproductive technology it seems difficult, if not impossible, to avoid making eugenic decisions. At the beginning of this paper I suggested that some argue against the reproductive technologies simply because they may be used for this purpose. I should like to turn this around, and suggest that eugenics does not really introduce new problems into the use of the technology. If we want to argue against using these means for eugenics, we need to argue that the means are undesirable on other grounds. There may be good arguments for such a view. This is obviously a subject for separate discussion, but the following points seem pertinent here.

The main argument for introducing reproductive technology has been to satisfy the desires of childless couples. It seems a small step from that to satisfying their desires for the sort of children they want, and we have looked at the question in terms of parents choosing to have children with enhanced genetic potential. But it might be held that it is not enough to look at questions such as these on an individual basis. There is a wider, social, question about the kind of society that will be produced by developing the technological control of reproduction, and by what criteria we are to assess its impact. It is to these questions relating to social change as a whole that the debate on reproductive technology should turn.

Notes

[1]   Cf. Robert L. Sinsheimer, 'The prospect of designed genetic change', in Adela S. Baer (ed), *Heredity and Society: Readings in Social Genetics*, Second edition, (Macmillan, New York, 1977) pp.436—443.

[2]   See R. Snowden & G.D. Mitchell, *The Artificial Family: A Consideration of Artificial Insemination By Donor* (Allen & Unwin, London, 1981) p.66.

[3]   See, e.g. Jon Beckwith, 'Recombinant DNA: does the fault lie within our genes', *Science For the People* (May—June, 1977) pp.14—17; B.F. Skinner, *Walden Two* (Macmillan, New York, 1948). This is not the only reason that might be given for a principle of non—interference, but it seems to me to be the most interesting. I am not going to discuss, in this paper, objections on the grounds that eugenics is 'flying in the face of nature' or destructive of what is 'truly human'.

[4]   J.M. Thoday, 'Geneticism and environmentalism' in A.H. Halsey (ed) *Heredity and Environment* (Methuen, London, 1977) p.30. N.B. His use of the term 'environmentalism' is different from mine, as it refers to a theory about the origin of characteristics.

[5]   See e.g. C.O. Carter, *Human Heredity* (Penguin, Harmondsworth, 1962) p.244. This argument is discussed and criticised by P.B. and J.S. Medawar in 'Eugenics', in *The Life Science: Current Ideas of Biology* (Wildwood House, London, 1977) p.56—65.

[6]   J.L. Mackie, *Ethics: Inventing Right and Wrong* (Penguin, Harmondsworth, 1977) Ch.6.

[7]   By A. Jones and W.F. Bodmer (Oxford University Press, London, 1974).

[8]   See Nancy S. Wexler, 'Will the circle be unbroken?: sterilizing the genetically impaired', in Aubrey Milunsky and George J. Annas (eds), *Genetics and the Law II* (Plenum Press, New York, 1980) pp.313—325. Nancy Wexler points out that to date suits

have been brought in the U.S. against medical personnel rather than parents, but take the view that the possibility of legal action against the latter cannot be ruled out. In English law there is apparently no action for wrongful life. In *McKay v Essex Health Authority* (1982) the Court of Appeal held that the common law did not recognise that a person has a cause of action for being allowed to be born deformed, and that the Congenital Disabilities (Civil Liability) Act 1976 had deprived any child of that cause of action.

[9]   I owe this point to Jonathan Glover.

[10]   The example comes from Gerald Leach, *The Biocrats*, revised edition (Penguin, Harmondsworth, 1972) p.120.

[11]   Estimated at 12% by June Goodfield in *Playing God: Genetic Engineering and the Manipulation of Life* (Hutchinson, London, 1977) p.66.

[12]   Bertrand Russell, 'Eugenics', in *Marriage and Morals* (Allen and Unwin, London, 1929) pp.200−214; P.B. and J.S. Medawar, *op.cit.*

[13]   Catherine Roberts, 'Positive eugenics' in James Rachels and Frank A. Tillman, *Philosophical Issues: A Contemporary Introduction* (Harper & Row, New York, 1972) pp.94−101.

[14]   ibid.

[15]   K.E. Tranoy, 'Asymmetries in ethics' *Inquiry*, 10 (1967) pp.351−372.

[16]   For discussion of whether the individual, the species, or the gene is the unit of selection see Richard Dawkins, *The Selfish Gene* (Oxford University Press, 1976).

[17]   Arno G. Motulsky, 'Governmental responsibilities in genetic diseases' in Aubrey Milunsky and George J. Annas (eds) *Genetics and the Law II* (Plenum Press, New York, 1980) pp.237−244.

[18]   P.D. Toon, 'Defining 'disease' − classification must be distinguished from evaluation', (*Journal of Medical Ethics*, 7, 1981) pp.197−201.

[19]   Ian Kennedy discussed this example in 'The Reith Lectures: Unmasking Medicine', *The Listener* Nov 6, 13, 20, 27, Dec 4, 11 (1980).

[20]   For suggestions as to which characteristics we should promote, see Gerald Leach, *op.cit.* Ch.4; C.O. Carter, *op.cit.* Ch.14.

[21]   In a radio discussion with Anthony Quinton.

[22]   Philippa Foot, 'The problem of abortion and the doctrine of the double effect', in James Rachels (ed), *Moral Problems*, second edition (Harper & Row, New York, 1975) pp.59−70.

[23]   For discussion of the sense in which the child can be harmed, if any, see Alexander Morgan Capron, 'The continuing wrong of "wrongful life"', in Arno G. Milunsky and George J. Annas (eds), *op.cit.* pp.881−93.

[24]   Jonathan Glover, *Causing Death and Saving Lives* (Penguin, Harmondsworth, 1977) p.148.

[25]   Cf. James Griffin, 'Is unhappiness morally more important than unhappiness?' *Philosophical Review* (1979) pp.47−55.

[26]   Catherine Roberts, *op.cit.* p.99.

[27]   Although this method has recently been attempted in Singapore., cf.

C.K. Chan, 'Eugenics on the rise: a report from Singapore' in *International Journal of Health Services*, Vol.15, No.4, 1985 pp.737—742.

[28]   Reported in the *Daily Express*, Friday July 23, 1982.

[29]   All of these problems are discussed in R. Snowden and G.D. Mitchell, *op.cit*.

[30]   ibid. pp.100ff.

[31]   Robert L. Sinsheimer, *op.cit*. p.442.

[32]   ibid.

[33]   Ruth F. Chadwick, 'Cloning', *Philosophy* 57 (1982), pp.201—209.

9 Transsexuals and werewolves: the ethical acceptability of the sex-change operation

HEATHER DRAPER

I   Introduction

The purpose of this paper is to reflect upon some of the ethical and philosophical problems which can be observed in the 'treatment' of transsexuals by the so called sex−change operation.

Transsexuals are people who believe themselves to be of the opposite gender to that which they overtly are. They should be distinguished from transvestites − who cross−dress − and homosexuals, who find members of the same sex sexually appealing. The transsexual may cross−dress and have sexual relationships with members of the same sex, but this is because he believes himself to be a female, or she believes herself to be a male, and they are therefore exhibiting normal behaviour within their own frame of reference. Transsexuals should be further distinguished from people who are physically bi−sexual. For example, in testicular feminisation, the subject is frequently very beautiful, with feminine traits and orientation; she has, however, the XY chromosomes of the male, a testicle in each groin, and a short vagina. Limited surgery may be required in order for her to function sexually as a female. Other instances of uncertain physical sexuality can be observed in Kleinfelter's Syndrome and Turner's Syndrome. It has been observed that these people rarely have a crisis of sexual indentity, nor do they request a sex−change operation [1].

Thus, transsexuals are people who believe that they are occupying the body of the wrong sex. They are sometimes referred to as "She−males" and "He−females". [2,3]

The sex−change operation is the surgical procedure which attempts to align the transsexual's physical appearance with their belief about their sexuality. Surgery to the genitalia is neither new, nor is it alien to western medicine. In the eighteenth century, the barbarities of castration, infibulation and clitoridectomy were not only accepted, but energetically sought by those 'suffering' from symptoms related to and arising out of the 'disease' of masturbation. More recently, operations were carried out on homosexuals to prevent them from satisfying their sexual appetites, and sex therapy suggested that some women should have their clitoris lowered towards the vagina to make it more accessible to men during intercourse.

The male sex—change operation includes the removal of the penis and testicles, which are replaced by a vagina. For the female operation, the breasts, genitalia and reproductive organs are removed and sometimes replaced with a false penis. Hormone treatment, which precedes both operations, continues afterwards.

The success of this operation is open to interpretation. The male—female operation is most acclaimed, though success can never be guaranteed and recovery is both slow and painful. The female to male operation is far more drastic, involving several separate major operations and, where a false penis is attached, numerous skin grafts which leave the abdomen very scarred. Once the penis is formed there is no certainty that it will function as a dispenser of urine, nor that it will even remain attached to the body! A distinction must be made, however, between the success of the actual surgery, and the achievement of the end for which surgery was merely a means. It is in the attainment of this goal that the operation should be judged. Success rates of as high as 70—80% were recorded when the operation was first tried; however, later, when long—term figures became available, success was estimated at nearer to 15%. [4] Owing to media coverage of the successful operations, and the desire of the subject of any failure to maintain their secret sexual identity, the public may have a distorted view of how sucessful the operation actually is. Failure is frequently marked by a return to a depressed and suicidal state. Due to these long—term figures, John Hopkins Hospital, the pioneers of the operation, declared in 1979 that they would no longer be performing it. Dr Jon K. Meyer, the hospital's psychiatrist—director, is quoted as announcing that a sex—change "...does not cure what is essentially a psychiatric disturbance". [5]

In so saying, Meyer highlighted two of the major philosophical and ethical issues; the identification of the nature of transsexuality, and the justification of allowing perplexed persons to be given an operation which (to quote Szasz) will merely turn them into "fake men and fake women".[6] It is possible that the operation is destined to failure because it can never fulfil the purpose for which it is designed; men and women can not be recreated in this way. Furthermore, the created organs do not function normally; for instance, the refashioned penis can not achieve an unaided erection, it is not especially sensitive, and orgasm is impossible.

Transsexualism tends to be a self—diagnosed complaint, and one almost impossible for a physician to confirm, or deny. In this case, clinical diagnosis centres on attempts to discount other possible causes (psychological, biological or otherwise) for the patient's beliefs and actions. Virtually all transsexuals have other identifiable psychiatric illnesses, and one quarter will have received psychiatric hospitalisation. Accordingly, it is difficult to determine whether transsexualism is part of an overall state of mental ill—health, or whether it is the patient's confusion about their gender which has resulted in other psychiatric symptoms.

In order to examine some of the problems involved, a comparison will be drawn between two methods of assessing the sex—change operation, both using analogies with other areas of modern medical practice. The first sees the sex—change operation in terms of another, unusual, request for permanent and mutilatory surgery, whilst the second analogy compares it to a request for cosmetic surgery.

II   The Werewolf Analogy

Suppose a man presents his G.P. with a request for help in disposing with parts of his body which are, he feels, preventing him from fulfilling his natural role as a werewolf. He claims that he is unable to continue with his life unless this help is given. The procedure which he is requesting is relatively simple, given current surgical technology. He would require alteration to his nose and ears, some hairy skin transplanting onto his forehead, (he already has an impressive beard), and also to the backs of his hands and feet. His toe and finger nails will also have to be altered, perhaps just split at the roots so that they resemble claws, and he will have to be referred to a dentist for a plate of appropriately shaped false teeth. The G.P., concerned for the patient, but keeping an open mind, asks how the man is sure that he will be happy with this surgery, if it can be arranged for him. The man promptly replies that he has considered himself to be a werewolf for as long as he can remember; likewise he has always been nocturnal, hence his job as a night watchman. He has also managed to live as a werewolf for the past five years, by using makeup and wigs to disguise his human form. He adds that he is serious in his suicide threat because he can no longer live as a human, using cosmetics to disguise himself, when underneath he is sure that he is a werewolf. The G.P. questions him about his aims in life and he replies that he just wants to be allowed to howl at the moon in peace, without being considered a freak.

The G.P. is determined not to ignore the patient's suffering, but is uncertain whether to refer him to a surgeon or to a psychiatrist, because his request seems to reflect the rantings of a disturbed mind. The case of the werewolf demands closer examination because it might follow that the grounds for refusing the werewolf his operation are applicable in the case of the transsexual.

This werewolf has several things in common with the transsexual. First, he is requesting surgery which is within the realms of possibility, and in this sense his request is a reasonable one. Second, he has a firm belief about his identity which is contrary to his appearance as it can be perceived by others, and this is substantiated by long years of successful imitation (transsexuals are frequently required to live as a women/man for at least two years before an operation will be considered.) Third, there is the threat of suicide which may also accompany requests for a sex—change. An examination of these common denominators highlights some of the ethical difficulties with the sex—change operation.

In the U.K., it is not possible for one to receive any N.H.S. surgery merely because one has requested it. Surgery is considered only where there is good reason to suppose that it is necessary and will succeed. The necessity of the surgery in this case is related to the way in which transsexuals and werewolves are to be categorised, and this will be discussed shortly. A distinction has already been made between the success of a surgical procedure and its outcome in relation to the purpose for which it was undertaken; hence the cliche, "the operation was a success but the patient died." Once the surgery is completed it will be too late for the transsexual or werewolf to change their mind; therefore all efforts must be made to ensure that fully informed consent is obtained. The success of the surgery will, however, depend largely upon how it is viewed by the patient. His/her impression of what the surgery can achieve, and what s/he believes their new role to be, will be influential. Accordingly, an examination of the beliefs of the transsexual and those of the werewolf, and how these are to be interpreted, is necessary.

In psychiatric medicine it is not uncommon for patients to make unrealistic claims about themselves or their experiences but one should not presume that these claims are fanciful. Occasionally, they have been substantiated after careful inquiry; for example, someone may in fact have been drilling holes into an adjoining wall, and this has literally driven the patient mad. The claims of the transsexual/werewolf are more difficult to substantiate. Regardless of how they have managed to live, and how vehemently they argue their case, to a large extent it remains a question of their word against the "fact" of their apparently normal body. This does not mean that it is impossible to label someone "transsexual", for this label merely describes their belief, in the same way that "homosexual" describes the action of one man in relation to another. What cannot be verified is the truth — value of the transsexual's claim. There is some similarity between the claims of the male transsexual and those of a woman with testicular feminisation (T.F.). The transsexual claims that despite his appearance he is really a woman; the woman with T.F. claims that, despite the evidence of her male XY chromosomes and hidden testes etc., she is neither a male, nor an hermaphrodite, but a woman with feminine desires, impulses and traits. The claims are very similar, yet the transsexual will have greater problems in convincing the doctors of his case than the woman, whose claim is supported by her sexual role since childhood and her birth certificate. This element of social acceptance is an important and influential one, showing that in many respects sexual identification is a social phenomenon which may or may not be supported by physiological facts. Hence it is possible to draw the important distinction between sex and gender, where sex is a medically verifiable 'fact' and gender is a socially constructed role. If the transsexual's claim is to be accepted, however, for consistency's sake so must the claims of the werewolf, other psychiatric patients and even 'normal' people who make claims about themselves which contradict the visible evidence.

Next, the importance of the suicide threat in medical decision making must be examined. Treatment which is likely to remove the threat of suicide has sometimes been seen as life — saving, enabling surgeons to condone sex — change operations on the grounds that they were necessary to save life. This is an inaccurate assessment for several reasons. First, there is no justification for giving a treatment life — saving status because of a suicide threat. Renal failure is life — threatening because, if dialysis is unavailable, it will cause the patient's death. Transsexualism will not result in death: if a sex — change operation is refused, the patient is the cause of his/her own death. This observation lends support to the view that the sex — change operation is attempting to relieve symptoms rather than their underlying cause, principally the patient's beliefs. Second, the operation is no guarantee of future happiness and mental stability. Whilst there is little documented evidence about suicide rates amongst transsexuals, Lothstien reports, anecdotally, that the suicide risk is slightly greater after the operation than before.[7] Finally, it should be noted that, since suicide is recognised as a sign of mental disturbance, the credibility of the transsexual's claim is shaken, giving grounds to suppose that s/he should be protected from his/her own self—destructive tendencies. Szasz, sadly and astutely, suggests that something is amiss when patients' demands are met as the result of a suicide threat. He makes particular reference to abortion, noting that doctors are more hostile to a cold and calculated request than to one which is accompanied by a histrionic suicide threat. Whilst suicide threats should be taken seriously and handled sympathetically, there is no good reason for them to influence a medical decision by creating the illusion of a life—threatening situation. A serious suicide threat generally indicates a disturbed and possibly irrational mind; but at the same time it might also be viewed as an indication of the sincerity of the transsexual's beliefs.

It could be argued that the werewolf analogy is inappropriate because whereas men and women, masculine and feminine, can be observed, and play a cruxial part in our society, there is no such thing as a werewolf. This objection fails to recognise the purpose of the analogy. In this context, what the patient actually believes is irrelevant; what is relevant is that he believes himself to be something which he overtly is not. Furthermore, werewolves do exist in the sense that they are characters in literature, like mermaids, unicorns and Hamlet, with whom we can, even as children, identify. The werewolf analogy undermines the validity of the sex—change operation by interpreting it as an attempt to solve psychosis by the mutilation of the genitalia. Accordingly, it can be argued that it is unethical for doctors to mutilate a healthy body in an operation which is so frequently unsuccessful, thereby exposing the patient to operative and post—operatve danger and pain unnecessarily.

III   The Cosmetic Analogy

It can be argued that the werewolf story is an inaccurate comparison with the sex—change operation, and that the feelings involved are more akin to those normally associated with requests for cosmetic surgery.

Suppose that this alternative scene is enacted. A man presents himself to his G.P. with a large and ugly birthmark on one side of his face. It is an obvious mark and not one which could be covered by make—up or a change of hairstyle. He is in a distressed state and threatens to commit suicide unless something can be done to rid his body of this mark, which he perceives as destroying his chances of any happiness. He considers himself to be socially inadequate, due to lack of confidence caused by people staring at him as though he is a freak. Once this mark is removed he is convinced that he will be able to lead a normal life.

On this occasion the G.P.'s decision seems more clear—cut. There are provisions for people to have cosmetic surgery financed by the N.H.S. provided that the blemish is causing sufficient mental or physical suffering, as it is in this case.

It can be argued that the male transsexual is in a similar position to this patient. He feels totally inadequate because, despite his good health, his body is blemished with testicles and a penis. He views his genitalia as an abnormality, just as the other man does his birthmark. The male transsexual wishes to be restored to the normality of a female body. *Prima facie*, this argument offers a challenge to the werewolf analogy, by altering the frame of reference from that of a mentally ill patient making a request for unnecessary mutilatory surgery, to that of a distressed patient making a perfectly understandable request for cosmetic surgery. However, it can be counter—argued that the cosmetic analogy is also inaccurate, because the birthmark is a tangible abnormality, whereas the transsexual's body appears to be normal. In this respect, a closer analogy would be one where the patient was requesting that two perfectly normal and not at all deformed ears were changed into a set which was not only scarred but also almost completely useless as hearing organs. Even if such a patient was suicidal there would be insufficient justification to risk the surgery. In this case, because there is nothing wrong with the ears, it could be argued that the patient's problems lie in the attitude towards his/her ears rather than in the ears themselves. It follows that it is this attitude towards the body which reqires the alteration rather than the body itself. The transsexual's problems may result from a similar mental rather than physical cause.

It does not necessarily follow from this that the sex—change operation should not be allowed. It has been proposed elsewhere in philosophy that people have a right to do as they like with themselves, provided that no—one else is harmed in the process. This is known as self—determination, and is evident in procedures where surgery is performed entirely for cosmetic effect, for instance to remove wrinkles, enlarge breasts, widen eyes, etc. This type of surgery is unnecessary, but nevertheless acceptable. It is unnecessary in the sense that the patient will come to no harm if it is left undone. Furthermore, the unchanged wrinkles, breasts, eyes etc., are not usually gross enough to be abnormal. Often they are actually normal for the age of the patient concerned. These operations are however an acceptable cosmetic commodity and are widely practised in affluent American circles. Despite there being no tangible deformity, such operations can be obtained as a matter of course.

Given this evidence, it is difficult to refuse the alterations requested by the transsexual. Patients receiving beautification surgery are similarly exposed to pain and the dangers of anaesthetics, for no sound clinical reason. There is, however, an important distinction between the two, and this can be found in the differing expectations of the patients. A woman may have her wrinkles removed so that she looks and feels younger, but she does not actually believe that she is younger, even though she may now lie about her age. Surgery knocks years off her appearance, not her life. The female transsexual does not consider that surgery will make her look like a man, she believes that it is the final step in a process which actually makes her into a man, she sees it as a literal sex—change, not as a cosmetic alteration to herself.

This analogy counters some of the objections to sex—change operations raised by the werewolf analogy. In cosmetic surgery, the patient's rights over their own body are accepted despite the risks and pain of exercising them. The analogy does not, however, tackle the problem of the transsexual's real identity because unless we actually consider that he is a woman (or vice versa), the sex–change operation cannot be considered to be merely cosmetic and at the same time to comply with the end for which the transsexual has requested it. It is seen by the transsexual as more than skin deep.

IV   The Role of Sexual Therapy

Before drawing any final conclusions, it is important to question and offer some explanation for the apparently unchallenged place of the sex–change operation within medical practice. How has an operation of such   drastic nature, and dubious success, become accepted as almost automatic   in the treatment of transsexuality? When transsexualism and the sex–change operation are seen within the framework of sex therapy as a whole, this question appears to be solved. A full discussion of sex therapy is not possible here; so the observations pertinent to transsexualism will be made without a full preamble.

Sex therapy, as practised by Masters and Johnson, suggests, both directly and indirectly, that there is a norm and standard of excellence which can be attained in human sexual performance. The norm is a heterosexual act, where both partners achieve orgasm without difficulty or unnecessary delay. Despite (so called) sexual liberation, men and women each have a definite role to perform, and can do so with varying degrees of competence. Sexual emancipation, especially in the case of women, was an overdue and applaudable social change; however, it seems that these two concepts of normality and competence have trapped sexuality within its own emancipation. Consequently, instead of couples discovering a norm which is right for themselves, they feel obliged to achieve a norm constructed by the therapists. Whilst sex therapy has claimed to have helped many, (exorcism also worked in its time), critics suggest that it is creating as many problems as it is apparently resolving, one of which is an inverted sexual prudety. Sexual demands are now insufficiently questioned because people do not wish to appear old fashioned or prudish. This 'anything goes' attitude has meant that the bizarre nature of the transsexual's claim has been overlooked because of its sexual content, whilst the werewolf's problems are seen in greater perspective, as both bizarre and ridiculous. Unfortunately, the sex–change operation may itself have arisen from the concept of normality. The male transsexual may be more feminine than the average woman, but he must further prove himself by being able to engage in 'normal' heterosexual activity. Due to the dictates of the sex therapists, this means that he must have a vagina. Here the transsexual's confusion between sex and gender is apparent. Viewing transsexuality as a sex rather than gender crisis, the solution has been sought in the mutilation of the sex organs, rather than in the gender norms where the problems of the transsexual may in fact have their ultimate roots.

V   Conclusion

It is now possible to draw some conclusions about the sex–change operation. Arguments against the operation consist of its poor record of success, the problems of defining and verifying the beliefs of the transsexual, and the apparent invalidity of the operation as highlighted by the werewolf analogy. Ultimately, reason dictates that transsexualism is a psychosis created by delusion, and there is some evidence in psychology to support this view. In favour of the operation there are the patient's rights to self-determination, which are supported by tolerance of non-essential cosmetic surgery. There is also the question of what is to be done with the transsexual if s/he is denied the surgery which s/he considers as his/her salvation, and which s/he knows to exist. Medical ethics must move within the realms of possibility; there is little point to a system which gives rise to impossible choices for doctors. It is unfair merely to state that something is unethical and then to leave doctors to cope with the consequences. It is they who will have to tell the transsexuals that nothing can be done to alleviate their mental suffering. Further, the extent to which doctors can intervene to protect patients from themselves, especially those who are mentally disturbed, needs more precise definition. The sex-change operation is irreversible and of a mutilatory nature. It is doomed to failure if the nature of the problem it intends to solve is psychological, rather than physical. Likewise, it must be accepted that the operation is not in fact a sex-change; the net result is not a new woman or man, it is just a mutilated man who chooses to call himself a woman or vice versa. In this important respect the credibility of the transsexual's claim never alters. S/he will, for instance, always remain a man purporting to be a woman, not because he is infertile, nor because he lacks reproductive organs or proper genitalia, but rather because he has no past as a woman and will always have been a male. It may be possible to change gender, but it is not possible to change sex.

To conclude, if the sex-change operation is to continue, and its very existence seems to have created an insatiable demand for it, then let us be realistic about what it is and what it can do. Today, people expect medical science not only to cure them, but also to give them what they want, as opposed to what they need. This popular misconception is partly caused by medicine interpreting health as "complete physical, mental and social well-being".[8] Accordingly, medicine has begun to expand into other fields in its attempts to cure the whole person, and doctors deal with abortion, birth control, cosmetic surgery, sex-change operations etc., which are not medical problems by true definition. There is a whole new spectrum of illness which should be labled 'social ailments' as they are not illnesses in the true sense of the word. At the present time a transsexual cannot be cured, but s/he can be mutilated and deceived into well-being; or else the operation will fail but "at least someone tried". Let us not be deceived, though; this is not medicine. The sheer logistics means that only doctors are qualified and competent to help, but the spirit and concept of the sex-change operation really means that it is beyond the scope of medicine. The problems of the transsexual, and those of the werewolf, cannot be solved by surgery; doctors have merely choosen to involve themselves in a hopeless situation where others could offer no solution. Some therapists and psychologists, like Lothstein, are now convinced that therapy does offer the transsexual some real hope in resolving his or her problems. The existence of the sex-change operation has meant that previously therapy was merely directed towards that end, and not exploited to the full. There is also the problem of those transsexuals who consider that the operation is their only solution because that is what they think they want. Some success is now recorded where therapy is conducted with an open view concerning its end; so perhaps soon sex—change operations will become just another sexual mutilation of a past era.

Notes

[1]   Sir Martin Roth, 'Transsexualism and the Sex Change Operation: a Contemporary Medico—Legal and Social Problem', *Medico—legal Journal*, 1981, 49:1, pp.5—19.
[2]   J.G. Raymond, *The Transsexual Empire: the making of the She—Male*, (Boston: Beacon Press, 1979.).
[3]   T.Szasz, *Sex: Facts, Frauds, and Follies*, (Oxford: Blackwell, 1980.)
[4]   Roth (1981).
[5]   *Medical World News*, 1979, 17 Sept, pp.17—19.
[6]   Szasz (1980), pp.87 & 90.
[7]   L. Lothstein, *Female to Male Transsexualism*, (Routledge & Kegan Paul, 1983).
[8]   The World Health Organisation definition.

10 Why is pornography offensive?

DAVID LINTON

As a working definition of "pornography," I shall use the one offered by the authors of the *Longford Report*.[1] They regard pornography as one particular type of explicit sexual material; and they quote, and presumably endorse, the view of Mrs Mary Miles, a child psychotherapist, that "Pornography treats of sexual practices divorced from any tender consideration for one's partner. Erotica deals with the pleasure and art of sexuality, but always in terms of a positive emotional relationship, whether transitory or lasting". The essential characteristic of pornography is the dehumanizing and degrading of sex, which it produces through its separation of sex from love; it "exploits and dehumanizes sex, so that human beings are treated as things and women in particular as sex objects".

Yet we treat people as things or mere means in many spheres of life — for example, when we deal with them as shop assistants or bus—conductors, or conversely as customers or passengers — and this is rarely considered objectionable. What would be morally objectionable would be a refusal to treat them in a more personal manner under relevant circumstances; but this is not an essential characteristic of either purveyors or consumers of pornography. The crux of the matter seems to be that people regard sex as something which ought to be *essentially* and *exclusively* a matter of personal relations; and whatever violates this rule causes offence.

A clue as to why this is so can be found in two recent books by the anthropologist Mary Douglas, *Purity and Danger* and *Natural Symbols*,[2] and especially in her account of the concept of 'dirt'. On pp. 47—8 of *Purity and Danger* Mrs Douglas notes that our ideas of dirt are much older than the knowledge of the existence of pathogenetic organisms on which our modern conceptions of hygiene are based. "If we can abstract pathogeneity and hygiene from our notion of dirt, we are left with the old definition of dirt as matter out of place... Where there is dirt there is system. Dirt is the by—product of a systematic ordering and classification of matter, insofar as ordering involves rejecting inappropriate elements." Soil in the garden, for example, is acceptable, but on the kitchen floor it is dirt, to be removed as soon as possible.

This categorizing tendency of the human mind is similarly relevant to the conception of the "unnatural" or "filthy" in human behaviour. What is held to be "unnatural" depends on what is held to be "normal". For example, "perversions" such as homosexuality may be associated with anomalous uses of the sexual organs and the anomalous direction of sexual desire; and this conception may have its origin in earlier views of the natural function of sex as being procreation.

Moreover, if Mrs Douglas is right, not only are anomalies always disturbing, but bodily anomalies are among the most disturbing of all. For she claims that there is a universal (or "natural") symbolic association between the human body and the structure of society and the structure of categories, which are a reflection of the social structure. An "anomalous" use of the body is thus a particularly severe threat to the integrity of the conceptual structure, unlike, for example, an anomalous use of language, as in poetry. Insofar as the conceptual structure of a society symbolically represents the social structure, a threat to its integrity is, in symbolic terms, a threat to the integrity of the society itself. (This is not as unlikely as it may appear at first glance, if we suppose the symbolic associations to operate through some unconscious mechanism.) On this analysis, the reason that sexual licence, personal unkemptness, etc., are often thought to pose a threat to the fabric or health of society is that they symbolise either freedom from rigid social constraint or rebellion against its demands or its existence. In *Natural Symbols* Mrs Douglas expounds this thesis, and provides evidence from comparisons of closely related societies, some of which tolerate bodily abandon, while others, which are organized on more rigid lines, react violently against its manifestations, presumably seeing it as a threat to their established order. Pornography, I believe, represents a similar symbolic threat to the social order, but at second hand, as it were: the *consumption* of pornography represents an indulgence of licentious sexual tastes which itself symbolizes freedom from or rebellion against social constraints.

But pornography is anomalous in another way as well. We associate sexual activities with private intimacies, which should properly take place only in private. This has to do with the strong association in our culture between physical and emotional intimacy, which demands a privacy of its own: we consider that the physical intimacy ought to be an expression of a pre—existing emotional relationship, that ought to be essentially and exclusively a matter of personal relations. The reaction against pornography on the ground that it exploits and dehumanises sex may thus owe some of its strength to the fact that it brings what is thought of as being an essentially private and personal aspect of life into the market—place, which is essentially public and impersonal. This sets up an ambiguity involving categories which are not merely distinct but are directly antithetical.

That it is the categorical ambiguity, and not the sexual explicitness alone, which offends is shown by the existence of "distancing mechanisms" by which the embarrassing or shameful potential of situations is disarmed. One such mechanism is the portrayal of sexual intimacy in a context of emotional relationships, however fleeting. That its efficacy is culture—relative is illustrated by the fact that pornography is defined in the People's Republic of China in such wide terms as to include mildly romantic writing and some Western Classical opera (*Longford Report*, pp. 112−3). Thus a large proportion of the world's population which has the concept of pornography seems to lack that of erotica, which in the West is generally regarded as acceptable.

A second distancing mechanism is "routinisation", or the gradual acceptance, through familiarity, of changes which are at first found shocking or outrageous. This may account for the gradual acceptance of an increase in the suggestiveness or exposure of fashions in dress, and of pin−ups in certain daily newspapers, and displays of nudity in the cinema and theatre and on the television screen. The routinisation tends to defuse both the sexually embarrassing nature of the material and its tendency to be perceived as a symbol of social indiscipline. The perception of novelty amounts to perception of an anomaly which has to be fitted into place in one's conceptual scheme of things in order that it should no longer trouble the mind; but repeated perceptions of what was originally a novelty allow the formation of a new category to which objects of such perceptions may be classified as belonging.

This is in keeping with Mrs Douglas's point that the potency of particular examples of licence and abandon as natural symbols is dependent on the degree to which they deviate from the norm. Secondly, it illustrates her thesis regarding the need to explain such potency by a comparative study of closely related cultures, displaying different degrees of rigidity in other matters. In such terms we may explain the anguish expressed by members of an older, more rigid culture, over the "decline in standards" regarding what is thought acceptable in the way of sexually explicit material, and their fear that this represents a decline in standards in general, and is a symptom, if not a cause, of "social decay".

Thirdly, there is the distancing mechanism of art. "Artistic merit" defuses the potential of sexual subject−matter to shock and embarrass, in a way akin to routinisation. By being portrayed in a manner instantly recognisable as "artistic" it can immediately be placed in a definite category as an art object, which is eminently acceptable to society, and hence not a symbol of decadence, although it may be situated historically as an example of the decadence of its period. The spectator may avoid the consciousness of his own sexuality and sense of sexual shame by diverting his attention from the subject−matter to the style and the purity of form, which constitute the marks by which the work can be categorized as a work of art in the first place.

Embarrassment is, however, often evinced by works whose artistic merit has been lauded by "authorities", but whose style is unfamiliar. When the artist and his style are already long−accepted a spectator may be confronted with the double embarrassment of his sense of sexual shame and the very fact of his own discomforture; but when the style is new and not yet assimilated, sexual subject−matter may produce both shame and moral outrage. Indeed, the novelty of the style itself may provoke an outcry, whatever the subject matter; it may be perceived as an offence against established standards and canons of good taste (which it well may be), and also an anomaly, out of place in the view of "respectable" art−lovers, and on a par with such threats to the social order as "perverted" or "unnatural" practices. If the artist's subject—matter is also sexual, the work will in all probality be condemned as pornographic.

But how does the artist's work itself, as opposed to how it is perceived, differ from that of the pornographer? Both might portray, for example, a female nude, and seek to idealize her physical sexuality and allure, as well as emphasizing her willing accessibility through her degree of exposure and her posture. But the artist is bound to portray his model according to artistic canons of gracefulness, whereas the "hard—core" pornographer may emphasize the gracelessness of her pose. Moreover, the artist will tend to bring to life the model and her sexual invitation entirely within the world of the picture, so that the viewer may see himself as the person to whom the invitation is extended only by fantasising himself into that world, which is clearly marked as distinct from the one he really inhabits by the artistic canons of style, perfection and grace. Where these are lacking, the invitation may appear to be aimed directly at the viewer, who can therefore no longer clearly separate the world of his sexual reaction from the real world, in which he might be embarrassed or ashamed. The apparent reality of the accessibility of the pornographic model may be further emphasized by the everyday nature and style of use of the photographic medium, and by a degree of commonplaceness combined with incongruity in the props and the situation in which she is placed.

This last factor may help to explain the outrage which met Manet's "Déjeuner sur l'herbe" when it was first displayed in public. A naked lady in a rustic scene would not have outraged anyone: nymphs "sporting unclad in woodland glades" have had a long history of acceptance in the tradition of Western art, even when showing an overt responsiveness to the courtship of a shepherd boy. What was so disturbing was to see a naked girl in the company of two fully dressed men, whose clothes suggested the social engagements of the bourgeoisie. This could have given offence in any, or all, of three ways. First, the anomaly of the mere juxtaposition could produce outrage. Secondly, her nudity and loose posture in a common—place setting, neither idyllic nor her own boudoir, intimates, like pornographic portrayals, her free sexual availability: there is no need to entice her to the bedroom, for she will disrobe at any time and place. Thirdly, the impersonality of her relations with the men, as portrayed in the painting, seems to indicate, as do the situations portrayed by pornographers, that she might be enjoyed sexually, not only at any time and place, but also by anybody, and in particular by the viewer.

It is the eventual acceptance of Manet's painting which now requires explanation. One main reason may be that it could be placed within the total corpus of works of an indubitably great artist *as an art object*, i.e. an object portraying a fantasy—world, so that many of those who initially responded with shock and indignation could have come to see that they had been frightened by a shadow of their own world cast by an image belonging to another. But even this would not have been possible in a society unused to the concept of change in standards and therefore unable to feel a little foolish at their initial response to having their standards mocked. Once the work had become accepted as an art object, and the symbols of bourgeois respectability, such as the style of dress of the men in the painting, had changed, its power to offend, associated with its portrayal of those symbols "out of context", would have diminished, and it could be seen as a joke, rather than a "real—world" symbol of threatening indiscipline. Established works of art, however obscene they may have seemed at the time of their publication, tend to be taken at face—value as art—works.

In another type of artistic or quasi—artistic fantasy, the portrayal of physical love may be presented as a celebration of an ideal other—worldly romantic love, which excuses all and dissolves all shame. Often artistically crass lyricism may be used to this effect, as in the film "Ryan's Daughter". The effectiveness of this device depends on involving the audience so totally that it cannot stand back and complain that the characters ought to do their love—making in private. Rather than distancing the scene portrayed from the audience, it must close the gap entirely, by drawing them into the reverie. The specialness, apartness and sanctity of the situation distinguishes it from the daily lives of the people caught up in it, and distances their reaction from their commonplace embarrassments and shame. Furthermore, because this particular form of abandon is sanctioned and approved by a large section of the public, it does not represent for them an illicit, socially dangerous, or threatening form of licence. It is similar to the Dionysiac type of orgies which are held at fixed dates of the religious calendar, in some primitive societies which at other times impose strict regulations on behaviour. For such peoples such occasions may serve, by their very contrast and cathartic effect, to emphasize and reinforce the strictly regulated mode of conduct of ordinary existence. So one may suppose it is with fantasies and reveries of romantic love.

One last distancing mechanism needs to be mentioned. This is the fact that in its proper place pornography need not, and often does not, cause offence. For example, "Soho is not quick to take offence. It is marvellously uninterested in what goes on next door. Restaurants where you could safely take your most prudish old aunt nestle up against brash new strip clubs". (Norman Shrapnel, "The porn market", *TLS*, 11th Feb. 1972, p.159). Again, in some homes nudity may be acceptable in the husband's "girlie" magazines, but not in pictures hung on the parlour walls, in what is defined as the wife's domain. Conversely, in more middle—class homes, what is acceptable in a book on Art, or a print or painting considered to have artistic merit, or even a Sunday colour supplement, may be unacceptable in the less chic pages of a "girlie" magazine. In both homes unsolicited literature offends by polluting the sanctity of the home, as do displays of nudity or physical intimacy on the television screen. It is such intrusion which arouses indignation and complaint; for most people, Shrapnel notes, as long as Soho keeps itself to itself it does not cause offence.

So it appears that what most people find offensive is not sexual explicitness, but sexual explicitness *out of place*; and that this is because it is perceived, consciously or unconsciously, as a symbolic threat to society's integrity. Indeed it symbolizes two kinds of threat. One is a blurring of categories and distinctions, leading to a situation in which roles, stations and duties are poorly defined, and there is a growing sense of "not belonging" and of passive revolt and non—cooperation. The other is active rebellion against the structure of society, and the institutions on which the production and maintenance of social welfare is thought to depend. Even from a "liberal", i.e. individualist and libertarian, point of view, these are genuine threats: if either reached a critical level, there might be insufficient people with the will or ability to perform the necessary tasks of industry and bureaucratic organization, or to teach the skills and transmit the will to use them to the rising generation. Another danger is that reaction to these forces might bring in an authoritarian and "illiberal" regime.

The general relaxation, in recent years, of the categories of moral thought has been in some respects desirable, inasmuch as it has replaced automatic emotional reactions with reactions based on more psychological understanding and self—awareness. But the under—valuing of the need of children and young people for some degree of authoritarian treatment in the home and in school has resulted in a degree of "anomie", or rootlessness and role—uncertainty, perhaps exacerbated by the conflict of modes of thought, and of the styles of dress and behaviour which symbolise them, between the permissively raised generation and their elders. This has sometimes produced severe alienation as well as anomie, and an increase in both passive and active rebellion. Further consequences for education in particular, in the opinion of many people, have been a decline in literacy and numeracy, and a devaluation both of academic study and of the skills necessary for the smooth running of society.

In so far as the rise in the production and consumption of pornography symbolizes this general rise in indiscipline many people will find it disturbing and offensive. But this does not mean that pornography itself causes the indiscipline; it may well relate to it as a symptom, symbolizing without aggravating the general development. And if these phenomena are merely symptomatic of deeper social problems, it would be senseless to try to suppress them without tackling the disease itself.[3] It would follow from this that it would be unnecessary, and perhaps undesirable, for the law to try to eliminate pornography as such; its main function, apart from the protection of minors, should be to protect people from the offence that is caused by their having obscene material thrust upon them, from the unwanted and unsought exposure to sexual or violent material, for example, on television, through the post and in public places.

Notes

[1] *Pornography: the Longford Report*, Coronet books, London, 1972, pp. 408—12.
[2] M. Douglas, *Purity and Danger*, Pelican books, 1970; *Natural Symbols*, Pelican books, 1973.
[3] For a discussion of this see J. Miller, "Censorship and the Limits of Permission" (British Academy Lecture, Oxford University Press, 1971).

11 Technology and psychotherapy

DAVID SMAIL

Psychotherapy covers an extremely broad and varied field of endeavour and takes many forms, which depend for their nature as much as anything upon the particular context in which they are set. It would for this reason almost certainly be misleading to attempt to develop any kind of general critique of psychotherapy, and I should acknowledge at the outset that what follows is based upon a personal knowledge of psychotherapy as it is practised within clinical psychology in the British National Health Service. This is, however, not an inconsiderable area: clinical psychologists, through their influence on theory if not through their practical intervention, have had an enormous impact on the way so—called 'neurotic' patients are 'treated', and it is also reasonable to expect that many of the issues relevant to psychological therapies in this area will be relevant also to psychotherapy more generally. I would however ask the reader to bear in mind that in any discussion of psychotherapy there are very few rules and many exceptions.

Despite attempts here and there to experiment with procedures of 'automated therapy', intrusion onto the psychotherapeutic scene of the paraphernalia of high technology has so far been insignificant. Patients do not sit behind banks of electronic machinery, nor, in established therapeutic practice, do they yet interact with computers. Though some therapists may make quite extensive use of audio or video recording, and a tiny minority use machinery for the registration of physiological variables, the setting in which psychotherapy takes place is, to all outward appearance at least, a matter of human intercourse. This is not to say, however, that psychotherapeutic theory and practice are not deeply imbued with technological values, and there is probably little doubt that, if these could be realized in the form of the reliable procedures and techniques which therapists dream about, financial and managerial interests would quickly establish a computerized therapy industry which would threaten to make the therapists themselves redundant. However, despite the fact that there is an enormous demand for psychotherapy and counselling for almost any conceivable kind of 'problem', and an already evident entrepreneurial interest in this potentially lucrative market, it would, for the time being anyway, be beyond the resources of money and ingenuity of even the most determined advertising campaign to persuade people that 'we have the technology' to cure their psychological ills automatically. On the other hand, the technological assumptions which underpin so much of psychotherapeutic thinking are widely and uncritically shared by both practitioners and patients, and it is upon such assumptions that I want to concentrate in this essay.

Although it is a statement which I am sure many psychotherapists would want to challenge, I can think of no major school of psychotherapy which does not view its task as finding out what is the matter with people and putting it right. In this way, therapy is seen as practical and empirical, and patients as, fundamentally, disordered mechanisms. Psychological symptoms or emotional disturbances are seen as signs that something has gone wrong *inside* the person which needs changing or adjusting (in his or her experience or 'cognitions', if not exactly in some physical structure). I have, of course, stated the matter crudely, but this essentially technological view of 'psychological disorder' and its 'treatment', though clothed in a wide variety of conceptual disguises, is what provides the justificatory rhetoric for all those therapeutic approaches which the average patient is likely to encounter.

Had I couched the preceding paragraph in language more sympathetic to the therapeutic enterprise it would be hard even to guess what objection I could have to it. Our culture is so saturated with technological values and ways of thought that to view an individual's disturbed behaviour as the projection of some kind of internal psychological problem seems not only reasonable, but the purest common sense, as does any attempt to ameliorate the disturbance by enquiring into and trying to change the internal psychology. It seemed obvious to the early psychoanalysts, for example, that the patient's discovery of *why* he or she reacted neurotically should, through the process of 'insight', lead more or less automatically to the resolution of the neurosis. Similarly, it seems obvious to the behaviourist that identification of the faulty conditioning process which has led to certain responses keeping company with the wrong kind of stimuli will indicate the correct programme of 're–conditioning' to put matters right. However, it seems to me that even in using terms like 'psychological disorder' (let alone 'neurosis' and 'treatment') we are being deceived by a technological mythology which threatens almost totally to obscure — perhaps indeed is *designed* to obscure — an understanding of the roots of human distress. Furthermore, the actual experience of psychotherapy teaches lessons about the nature of such distress which quite directly contradict the technological assumptions which therapists, despite their continuous exposure to those lessons, persist in trying to buttress.

There is a vast literature on research into psychotherapy which testifies to a tireless search on the part of practitioners and interested academics for potent therapeutic factors and techniques which can be relied upon to 'work' predictably (see Garfield, 1978, for just one representative tome).

There is some recognition in this literature that *human* factors may be of importance in producing those benefits which do sometimes seem to follow from psychotherapeutic involvement, but these are regarded with some suspicion. It is, for example, quite often grudgingly admitted that the kind of relationship which develops between patient and therapist, the kind of people they are and the way they affect each other may have significance for the final outcome, but such factors as these are typically characterized as 'non−specific', and seem disappointingly unlike the hoped−for discoveries of *technical* procedures which would more reliably deliver the goods. The usual conclusion for researchers in this predicament to reach is that, interesting though human factors may be, what is really needed is a much finer differentiation and measurement of the 'variables' which enter the therapeutic equation; we have not, so it seems, been able to elaborate a technical understanding of therapy only because our methodological tools have not yet been made sufficiently precise − but that day, it is confidently assumed, will not be long in dawning. Since we have yet to develop and apply a suitably sophisticated research methodology to the phenomena of psychotherapy, we need not worry over−much, it is argued, that so far nobody has really been able to demonstrate to the satisfaction even of the relatively gullible that psychotherapy actually does any good.

However, not everyone has been so impervious to the lessons which psychotherapeutic experience and research *do* seem to teach. There have always been strong therapeutic voices raised against the mechanist or technical approaches to therapy, though they have always been in a minority and even then have, in my view (which is elaborated in greater detail in Smail, 1978, 1984), for the most part failed to take account of the implications of their own criticisms. For example, Jung's objections to the impersonality and sexual reductionism of Freudian psychoanalysis did not preserve him from a kind of occult mechanism (e.g. in the operation of archetypes, the compensatory dynamics of dimensions of personality, etc.) and certainly not from a highly individualizing approach to psychological distress − for Jung it is the 'inner' world which is almost exclusively of interest. Suttie's (1960) correct identification of love as the single most potent therapeutic influence did not lead him closely to trace the implications of contracting to dispense it *professionally*, and Rogers' (1957) 'humanistic' elaboration of the necessary and sufficient conditions of therapeutic 'personality change' led to one of the most mechanistically ambitious programmes yet to isolate, train, and even build into computers, the requisite therapeutic 'skills' (such as warmth, empathy and genuineness) with whose discovery he is credited. In recent years there have been some excellent books written about psychotherapy which demonstrate its nature very clearly, testifying even movingly to the human nature of the undertaking (e.g. Lomas, 1981, Hobson, 1985) but it is perhaps only Thomas Szasz (1973) who has really challenged the most fundamental technological assumption of psychotherapy, that through tinkering in one way or another with people's psychological mechanisms you can alter their experience of the world.

For the fact we have to face, and which is in my view revealed directly by the experience of psychotherapy as well as indirectly by the inconclusive nature of a massive research effort, is that human beings are not machines and their distress (their 'emotional problems' and their 'symptoms') is not caused by 'dysfunction' located, if not physically inside their own individual skins, then psychologically inside their own individual 'heads'. Why is it that this fact seems so hard for us to assimilate?

I could not possibly try to provide anything like a comprehensive answer to this question within the scope of this essay, and even the very partial answer which I shall attempt contains a strong component of conjecture. I realize, of course, that I have as yet indicated no evidence to suggest that a technological approach to emotional distress, i.e. an approach which concentrates on the technical adjustment of an essentially mechanical fault, is not in principle viable. I shall attempt shortly to make good this omission, but first I should like briefly to suggest what one or two of the factors may be which contribute to our reluctance to abandon a technological stance.

Technology is magic, and as such represents an investment Western society has made in the fulfilment of its dreams over at least the past five centuries. The expectation expressed by Francis Bacon that scientific enterprise would bring to concrete realization what formerly had been the merely empty claims of magicians has in large part proved more than justified. As Keith Thomas (1973) writes:

> Francis Bacon listed as *desiderata* the prolongation of life, the restitution of youth, the curing of incurable diseases, the mitigation of pain, the speeding up of natural processes, the discovery of new sources of food, the control of the weather, and the enhancement of the pleasures of the senses. He wanted divination put on a natural basis so that it would be possible to make rational predictions of the weather, the harvest, and the epidemics of each year. His aspirations were the same as those of the astrologers, the magicians and the alchemists, even if the methods he envisaged were different.

Such achievements, however, have almost all been made in the fields of the physical sciences and the technologies which they have spawned. Scientific activity is no doubt informed by an age—old yearning for control of the environment and personal enrichment, but its methods are by now significantly different, and infinitely more effective, than those of the magicians and alchemists who first attempted to realize them. In the psychological field, however, this is far less obviously the case, and Thomas's fascinating book is full of striking parallels between the practices and claims of, in particular, sixteenth and seventeenth century 'cunning men' and astrologers and those of modern psychologists and psychotherapists. In this way, the procedures of therapeutic psychology, while employing a 'scientific' vocabulary and laying claim to technical efficacy, seem to have evolved but little from those of would—be healers of former times (there are also uncanny resemblances between, for example, present day procedures of psychometric testing and the descriptions Thomas gives of the construction and functions of seventeenth—century horoscopes). The expectation that someone can be significantly changed through weekly (or even daily) visits to a therapist in which all that takes place is talk — no matter how emotionally charged — does indeed seem on the face of it no more or less credible than a belief that a person may be psychically transformed through exposure to a witch's incantation.

We have always been and no doubt will always be credulous in such matters, and for this reason alone psychotherapists may rely upon a plentiful clientele. (I should hasten to say at this point that I do not in any way wish to accuse therapists of charlatanism — we are all subject to the illusions of our culture. Psychotherapists are as genuinely convinced of the efficacy of their procedures as astrologers were of theirs.) We are also, of course, extremely vulnerable to the play of interest. In an age when technology enjoys practically unassailable prestige, it is scarcely surprising that psychotherapists should wish to associate themselves with its benefits, and indeed it is becoming increasingly difficult for ordinary people to think of themselves as anything other than complicated machines (e.g. as bundles of programmable 'skills'). In this way both our history (the impetus of our culture) and our interest push us in the direction of a technological approach to our emotional distress.

There seem to be times, and this is one of them, when what we take to be true is determined almost solely by our interests, and our standards of reality are set by our collective immersion in a kind of communal fantasy. The 'progress' over the last forty years or so of psychotherapy reflects this process, in my view, particularly clearly. Psychotherapy, along with several other aspects of applied psychology, established itself as a 'credible' discipline largely on the grounds of a willingness on the part of its practitioners to submit their activities to a (largely self—initiated) *scientific* scrutiny. Although, of course, Freud's claim to a fair hearing for psychoanalysis was always made, and not by any means entirely unjustifiably, on the grounds of its scientific status, nevertheless, until about the end of the Second World War, procedures of psychotherapy had relied for their acceptance, whether knowingly or not, on a very distinct aura of mystique. This was successfully punctured by the criticism, and indeed derision, of a number of advocates of the self—consciously hard—headed 'scientific' school of psychology, particularly behaviourists such as H.J. Eysenck. The real take—off point for professional psychotherapy, and a dramatic increase in the extent to which it became available to 'ordinary' people, came with the ensuing rush of therapists of all kinds and persuasions to demonstrate their effectiveness according to criteria generally accepted as 'scientific'. Many schools of psychotherapy, indeed, actually grew out of or became established through this burst of activity. However, the most important and tangible result of this interest in and proliferation of psychological therapies was in fact not a scientifically satisfactory demonstration that all or any of them 'worked', but rather that a whole new profession, or collection of professions, of psychotherapists and counsellors became firmly established and meshed in with a network of interests — their own, their clients', and those of the wider society.

Psychotherapists in the 1980s are in fact no nearer demonstrating the efficacy of any form of psychological treatment than they were in the 1940s, and, interestingly, they now for the most part seem far less interested in doing so. The reason for this lies in there being no longer any pressing reason to do so: psychotherapy and counselling — 'therapeutic skills' and 'counselling skills' — have become an accepted feature of our technological landscape and their efficacy, though unproven, is taken as beyond question. What in fact seems to have happened is that psychotherapy was borne along on a sustained wave of scientific/technical *rhetoric* to a point where it became established within the structure of technological 'credibility' which pervades our society, and at this point the scientific rhetoric — that which purported to enquire into the *truth* of claims made — was quietly dropped. To ask *scientific* questions about the efficacy of psychotherapy or the accuracy of theoretical claims made now invites one's being regarded by practitioners with a certain confusion and puzzlement, as if one is committing an embarrassing anachronism or even offending against an unspoken rule of etiquette. For now a new rhetoric has become established, a semi–professional, semi–managerial rhetoric which *assumes* or tacitly *asserts* the efficacy of therapeutic 'techniques', and concerns itself only with questions of their suitability for particular cases or their applicability to particularly interesting problems of social engineering.

This predicament of psychotherapy, if one may be permitted to see it as such, seems to me to do no more than reflect a set of attitudes and beliefs which are deeply and pervasively rooted in our culture. We are gripped by a dream of technological power and success which will allow its dominance to be challenged by no amount of evidence of its fallibility. Too much is at stake, too many interests are involved, too many jobs as well as too many illusions depend upon the validity of our faith in technology for us to be able epistemologically or even morally to be able to challenge it.

And yet it is challenged, over and over again, in the experience of ordinary people, among them precisely those people who present themselves to professional helpers for some kind of treatment or therapy for their emotional distress. For it is not only the voluminous research which has been conducted into the outcome of psychotherapy which suggests that in fact, in the way that we want it to, it does not 'work' — it is also the experience of those who submit themselves to its procedures. This is not to say that psychotherapy is not valuable or of benefit to people, nor even that it is not a useful and necessary resource in modern society (I shall turn presently to a consideration of its positive contribution), but only that it does not cure people of what they expect to be cured of or change them in ways they hope to be changed.

Psychotherapy does not 'work' in this way precisely because human beings are not machines. Technological procedures of fault finding and repair are entirely appropriate to mechanical systems, but, as any gardener knows, other methods of care and cultivation are necessary for organic structures. Only a deeply rooted dogma of individualistic mechanism permits us, against all the evidence, to persist in seeing the pain and despair our society inflicts on people as the result of some kind of personal 'dysfunction' inside them. It is, I am sure, not without significance that adherence to that dogma also permits us, or some of us, to continue the 'pursuit of happiness' enjoined upon us by our culture without worrying too much about the casualties created along the way. After all, broken–down machines can be mended.

Despite the fact that in the heyday of its 'scientific' rhetoric psychology used to refer to people as 'organisms', psychologists and psychotherapists have always been too fascinated by the machine model to take seriously the organic nature of human psychological life and experience. The mechanist analogies to be drawn from medicine and surgery, and more recently from electronic and computer technology, have always made it almost impossible to think of people's 'symptoms' and 'problems' as anything other than the result of the malfunction of bits and pieces of 'psychic' or 'cognitive' machinery inside them; bits and pieces, furthermore, which can at least in principle be excised and replaced. While it is true that we continue, despite evidence to the contrary (e.g. Illich, 1977), to persuade ourselves of the soundness of the fields from which these analogies are drawn, there are still psychological aspects of our conduct which are so recalcitrant to mechanical forms of interference that we would not dream of trying to treat them in the way that we do our 'mental disorders'. For example, we seem to have a clear tacit understanding that the acquisition of language or of practical abilities like swimming or bicycle riding is not something which can be implanted by experts (whether magicians, surgeons, hypnotists, etc.) nor indeed something which can be simply erased or replaced. These are things which have to be learned by the subject, often painfully, and once learned cannot but for exceptional circumstances be forgotten. One cannot cure people of knowing how to swim, and though a native speaker of English can certainly learn Chinese, knowledge of the latter cannot replace or be swapped for knowledge of the former — one cannot eject the English programme and pop in the Chinese one. And yet this is precisely what we expect to be able to do with whatever structures of learning and experience underlie the 'symptoms' of emotional distress. Through 'insight', or 'corrective emotional experiences', 'catharsis', 'abreaction', 'interpretation', 'positive self—talk', 're—conditioning', 'systematic desensitization' and so on and on, we nurture the illusion that somehow we can clean up the psychic machinery so that it will respond in an 'adjusted' way to a world which is, presumably, taken as simply given, 'there', and to be adjusted to.

But human beings are embodied organisms who are located in a world which is as much the creation of social praxis as the creation of God or Nature. Despite frequent ritual acknowledgement of the philosophical problems presented by psycho—physical dualism (again a feature of scientific rhetoric never really taken seriously), psychologists and psychotherapists have always found it difficult to escape from conceiving of people as anything but, on the one hand, minds inhabiting bodies, or, on the other, *only* bodies or *only* minds. Thus behaviourists take up a precarious posture on the 'physicalist' horn of the dualist dilemma, while 'dynamicists', humanists and some 'cognitivists' still cling to the 'mentalist' horn; uncomfortably trying to get a purchase on both are those who are willing to envisage some form of philosophical hybrid like 'cognitive behaviourism'. We do not seem to be able to see that mind *is* body (and body mind), and that the pain and suffering we feel as somehow inside us are inflicted through our *bodily* relation to a world from which we can only *pretend* to escape. Our 'psychic experience' has been, as it were, inscribed upon our living tissue, just as has our knowledge of a mother—tongue. Since it has been inscribed *by* something, or perhaps better *through* our relation to an inescapable world, it makes little sense to try to treat our distress by rummaging around inside our bodies or, by the use of drugs for instance, blunting our awareness of the world we inhabit by interfering with the chemistry of our perception of it. Nor does it make any sense to try to alter the way we see or think about things through the use of some form of 'positive' self—suggestion. It is only the legacy of magic in our thinking which permits us seriously to entertain the idea that by some kind of process of insight or self—instruction we can change our experience of a reality which is none the less real for having been created in part by us. You cannot learn to speak Chinese or to play the piano merely by thinking about it or *imagining* yourself able so to do. Even though he appears still to believe in the possibility of a relative freedom from the past which I am unable to accept, Roy Schafer (1976) approaches the problem of 'cure' with unusual honesty:

> ... while the past may be partially re−experienced, reviewed, and altered through reinterpretation, it cannot be replaced: a truly cold mother, a savage or seductive father, a dead sibling, the consequences of a predominant repressed fantasy, years of stunted growth and emotional withdrawal, and so forth, cannot be wiped out by analysis, even though their hampering and painful effects may be greatly mitigated, and the analysand freed to make another, partly different and more successful try at adaptation. The analysand whose analysis has been benignly influential retains apprehensions, vulnerability, and characteristic inclinations toward certain infantile, self−crippling solutions, however reduced these may be in influence and however counterbalanced by strengthened adaptiveness.

A person acquires his or her experience as a plant acquires the structures of its growth. One may strive to correct the conditions of an environment which leads to, say, a tree's weak or stunted growth, its withering from lack of shade or moisture, its leaning before a prevailing wind, but one cannot (sensibly) correct such eventualities merely by concentrating on the body of the organism itself. The kinds of misery which lead people to seek psychotherapy would best be prevented through the development of a society in which values of care and justice were central. This, of course, is Utopian, and one can envisage no society from which confusion and cruelty will be absent, but at least we could stop pretending that the ravages they cause can be put right by technical intervention after the damage has been done. Even the most ruthless factory farmer has a more accurate knowledge of the importance of the *conditions* in which his produce is raised.

If psychotherapy 'worked' technically, in the way we would like it to, somehow to change the conclusions people draw (and act upon) from the painful lessons their embodied experience of the world has taught them, it would indeed constitute a magical panacea, a blessing of unimaginable proportions. But only in the right hands. For if what people feel and think and do were really so amenable to the technical 'skills' of therapists and counsellors, one need have little doubt that the interests of power would quickly conceive of ways in which such technical mastery could be put to 'good' effect. In this way the tough resistance to change of our organically rooted experience may be a blessing in disguise, and perhaps even lend a kind of positive significance to our suffering. Even if we cannot realize the dream of being able to soothe away the scars of a wretched childhood, nor can we all that readily persuade those who have been through its mill that our social order is based on the values of fairness and sweet reason. Psychological distress is a tangible reproof for the way we treat each other.

What psychotherapists do, and what they achieve with individual patients, need of course bear no resemblance to their rhetorical claims or theoretical fantasies, and the fact that we do not change or cure our patients in the way both we and they would like need not be all that dismaying. For psychotherapy can legitimately have other purposes than the technical adjustment of people to a society uncritically accepted as given, and it is probably the case that in fact, whatever their conscious intentions, psychotherapists do very often render services of great value to their clients. For example, psychotherapy can clarify the mystery which surrounds people's experience of distress, and the 'therapeutic relationship' can be one from which people derive great comfort, as well perhaps as the encouragement to confront difficulties from which previously they had shied away.

The technological rhetoric of psychotherapy buttresses a society based on exploitation, since it promises painlessly to heal emotional injuries incurred by people in their struggle for 'happiness', but it contrasts strongly with what is often the 'demystifiying' *effect* of psychotherapy, which is to clarify for people that their distress arises not from some kind of *personal* fault or inadequacy, but rather from their having been situated in an unavoidably difficult set of circumstances. Very few psychotherapists trace the evils that beset their patients to the economic and political structure of our society, but even so most will be concerned to guide their patients gently and sympathetically to an appreciation that their 'problem' is not one of personal moral failure, but one which *in fact* stems from their embeddedness in a world outside their own heads. (The irony, of course, is that therapeutically this is a largely unconscious process, most therapists happily believing that what they are engaged in is a technical procedure of individual adjustment.) Very few people become psychotherapists who are not fundamentally affiliative towards their fellow creatures and hence tolerant of their sins and compassionate towards their pain. Because of this, patients are likely to find considerable comfort from their relations with their therapist, and indeed for many their experience of therapy may constitute the first time in their lives that they have been able to command the attention of a truly concerned and competent human being who is willing, even if only in circumstances strictly limited by professional boundaries, to place his or her resources at their disposal.

If there is no possibility of curing, there is no shame in comforting. Psychotherapy, whatever hopes and claims are expressed for it, will never cure the emotional injuries which we inflict upon each other. Avoidance of such psychological damage will not be achieved by technological means, but only through our developing a society in which we treat each other with greater care and kindness.

References

Garfield, S.L. 1978, *Handbook of Psychotherapy and Behaviour Change*, New York: Wiley.
Hobson, R.F. 1985, *Forms of Feeling. The Heart of Psychotherapy.*, London: Tavistock Publications.
Illich, I. 1977, *Limits to Medicine*, Harmondsworth: Penguin Books.
Lomas, P. 1981, *The Case for a Personal Psychotherapy*, Oxford: Oxford University Press.

Rogers, C.R. 1957, The necessary and sufficient conditions of therapeutic personality change, *Journal of consulting Psychology*, 21, 95—103.

Schafer, R. 1976, *A New Language for Psychoanalysis*, Yale University Press.

Smail, D. 1978, *Psychotherapy. A Personal Approach.*, London: Dent.

Smail, D. 1984, *Illusion and Reality. The Meaning of Anxiety.*, London: Dent.

Suttie, I.D. 1960, *The Origins of Love and Hate*, Harmondsworth: Penguin Books.

Szasz, T. 1973, *The Myth of Psychotherapy* Oxford: Oxford University Press.

Thomas, K. 1973, *Religion and the Decline of Magic*, Harmondsworth: Penguin Books.